Nutrition Applications Workbook

Second Edition

Thomas W. Castonguay, Ph.D.
University of Maryland at College Park

Chengshun Fang
University of Delaware

Eugene J. Fenster
Longview Community College

Jill S. Golden, M.S., R.D.
Orange Coast College

Eleanor B. Huang, M.S., R.D.
Orange Coast College

Lorrie Miller Kohler
Minneapolis Community and Technical College

Ruth L. Kuzmanic
Naperville Central High School

Elaine M. Long, Ph.D., R.D., L.D.
Boise State University

Judith S. Matheisz
Erie Community College

Gail Meinhold, M.S., R.D.
Orange Coast College

Mithia Mukutmoni, Ph.D.
Sierra College

J. A. Thomson
University of Waterloo

Julian H. Williford, Jr.
Bowling Green State University

With contributions by Jessica Castonguay

THOMSON
WADSWORTH

Australia • Canada • Mexico • Singapore • Spain • United Kingdom • United States

Printed in the United States of America
4 5 6 7 09 08 07

Printer: Thomson/West

ISBN-13: 978-0-495-01184-2
ISBN-10: 0-495-01184-3

For more information about our products, contact us at:
Thomson Learning Academic Resource Center
1-800-423-0563

For permission to use material from this text or product, submit a request online at **http://www.thomsonrights.com.**
Any additional questions about permissions can be submitted by email to **thomsonrights@thomson.com.**

Thomson Higher Education
10 Davis Drive
Belmont, CA 94002-3098
USA

Asia (including India)
Thomson Learning
5 Shenton Way
#01-01 UIC Building
Singapore 068808

Australia/New Zealand
Thomson Learning Australia
102 Dodds Street
Southbank, Victoria 3006
Australia

Canada
Thomson Nelson
1120 Birchmount Road
Toronto, Ontario M1K 5G4
Canada

UK/Europe/Middle East/Africa
Thomson Learning
High Holborn House
50–51 Bedford Row
London WC1R 4LR
United Kingdom

Latin America
Thomson Learning
Seneca, 53
Colonia Polanco
11560 Mexico
D.F. Mexico

Spain (including Portugal)
Thomson Paraninfo
Calle Magallanes, 25
28015 Madrid, Spain

CONTENTS

INTRODUCTION FOR INSTRUCTORS

Welcome to Wadsworth's *Nutrition Applications Workbook*, and thank you for choosing our nutrition texts for your course! The purpose of this supplement is to provide you with a selection of activities, worksheets, and project assignments that will: (1) help you prepare for your course, and (2) engage and challenge your students by encouraging them to apply what they are learning about nutrition in practical and personal contexts. A wide range of assignments covering similar topics are provided to allow you the freedom to choose those that best fit your course. To assist you in identifying the most useful materials, the table of contents includes brief descriptions for each item or set of items, and guides to the case studies and diet analysis activities organize them topically.

This workbook is divided into two parts:

Case Studies: Twenty case studies illustrating the impact of nutrition on health, covering individual nutrients, weight management, food safety, and many other topics. A worksheet with discussion questions follows each case study.

Diet Analysis Projects and Activities: Six sets of diet analysis activities with worksheets are included. Many of these exercises are entirely self-contained and may be completed using *Diet Analysis Plus* reports which are printed in this workbook, while others require students to keep a record of their own intake and analyze it using either *Diet Analysis Plus* or hand calculations with a food composition table. Detailed instructions for recording daily intake and for preparing several written assignments based on *Diet Analysis Plus* reports are also provided.

We would like to point out several features **new to this edition** of the workbook:

- The exercises from the first edition's Module B have been newly revised and expanded to include more *Diet Analysis Plus* printouts, both seven- and three-day personal diet analyses, a new recipe modification exercise, and a new "Carbohydrate Tick Tack Toe" in-class activity. They have been reorganized into Modules B and C.
- New "1-Week Diet Analysis: Actual vs. Model Diet" assignment includes both analysis of the student's diet and the planning and refining of a more healthful diet using *Diet Analysis Plus*.
- New "Healthy Heart Analysis" guides assessment of chronic disease risk in several areas using Internet resources.
- New "Snack Assignment: The *Virtual Snack Shop*" challenges students to create a unique snack and then demonstrate that it provides 30% of the recommended intake for a target vitamin and mineral.
- New "Diet Project: Comparison of Two Distinct Diets" is a more sophisticated assignment appropriate for a nutrition course taken for science credit. Students recruit volunteers to provide diet histories, which the student can then analyze and compare/contrast in a report.
- New "2005 *Dietary Guidelines for Americans* Evaluation" utilizes the recently-published guidelines as a basis for consideration of personal diet and lifestyle.
- The "Diet Self-Study Exercises" and "Thinking Through My Diet" modules by Sharon Rady Rolfes do not appear in this edition because of space limitations, but they will be available for download from the on-line Wadsworth Nutrition Resource Center at http://www.nutrition.wadsworth.com.

We hope you find this workbook a useful resource, and that your students' learning is enhanced as they work through these assignments.

CASE STUDIES CORRELATION GUIDE

Case Study Assignment Key: Major Topics for Which Case Studies Are Appropriate

Case #1	Factors influencing food choices	Case #11	Fat-soluble vitamins
Case #2	Diet planning strategies	Case #12	Major minerals
Case #3	Digestion and absorption	Case #13	Trace minerals
Case #4	Carbohydrates	Case #14	Physical fitness and nutrition
Case #5	Fats	Case #15	Nutrition for pregnancy/lactation
Case #6	Protein	Case #16	Nutrition for children and teenagers
Case #7	Metabolism	Case #17	Nutrition for adults and the elderly
Case #8	Energy balance / body composition	Case #18	Relationship of diet to health
Case #9	Weight management	Case #19	Food safety
Case #10	Water-soluble vitamins	Case #20	Food and agricultural technology

Index of Minor Topics Covered in Case Studies

Case Study #1: Cultural Differences and Nutrition

Beth is a computer analyst from New York City. She is 25 years old, 5'5" tall, and, at 160 pounds, she is overweight. Beth's fast-paced day typically starts at 7 a.m. when she gets up for work. She is always in a rush in the mornings, and barely has time to grab a quick breakfast of an instant drink or breakfast bar and a cup of coffee with cream that she consumes on the subway on her way to work.

Beth works long hours at her job and has a heavy workload that keeps her tied to her desk. Around 1 p.m. every day she takes enough time to grab lunch at the deli next door. She is in the habit of getting the same foods every day, generally a prepackaged corned beef or pastrami hoagie with the works: lettuce, tomato, onion, mayo, mustard, and American cheese accompanied by a small bag of plain potato chips. Beth usually orders a 20-oz. cola, and when she is feeling really stressed by a difficult work situation, she picks up a single serving hot apple pie to eat at her desk. Apple pie is a favorite comfort food and reminds her of happy times and special family meals.

By the time Beth gets done with work it is 8:00 p.m. and she walks to the subway station one block from her office and heads home. On the subway she decides she will have her favorite quick meal for dinner, a frozen "homestyle" fried chicken with garlic mashed potatoes and a side of creamed corn for dinner. She enjoys this meal and finds the mashed-potatoes to be very comforting after a long, tense day at work.

Beth eats most of her meals as takeout food or frozen meals out of the grocery store. She is generally too tired to cook, plus she likes the taste of convenient meals. Her freezer is always full of frozen pizzas, large-portion heat-and-serve meals, and frozen burritos. Each evening she enjoys an hour or two of television, and reads over some papers from work before fixing a bowl of Rocky Road ice cream and heading to bed.

One day her company holds a blood pressure screening for all of its employees and Beth decides to participate. Beth is shocked to find out that at age 25 she has elevated blood pressure. This news puts Beth in a reflective mood. Lately she has been noticing that she is gaining weight, and lacks some of the energy she had in college. Her father died of a heart attack two years ago at age 58, and Beth decides she should make some changes in her life before she follows in her father's footsteps to an early grave.

In another part of the world, Anna is busy with her daily routine. Anna works in her family's produce store right outside of Prague, Czech Republic. She is also 25 years old and 5'5" tall. Anna weighs 150 pounds and like Beth, Anna is overweight. Anna's day typically starts around 7 a.m. Anna's breakfast is leisurely and consists of a small Kaiser roll, a slice of ham, a small tomato, orange juice, and a cup of coffee with cream. Anna likes to start the day eating a tomato and drinking orange juice because she knows that fruits and vegetables are healthy choices, and she always has a supply of tomatoes from the family produce stand.

After breakfast, Anna heads downstairs to work. The family produce shop is in the downstairs portion of their house. While at work Anna's brother carries the large boxes of produce into the store and stocks the shelves. Anna sits in a chair at the register and rings up sales. The store closes at 1 p.m. for lunch. In the Czech Republic foods are generally prepared and eaten in the home, and lunch is the largest meal of the day. Anna and her family members enjoy a lunchtime meal of soup, usually potato soup, plus roasted meat, such as pork, boiled cabbage and dumplings. She often eats bread with her meal such as rye or pumpernickel with butter.

After lunch, the store then reopens until 7 p.m., when Anna returns upstairs for dinner. For dinner, Anna often eats a small sandwich of sausage on a bun. With her evening meal she often has a glass of beer. Anna is generally in bed by 10:30 p.m., and tries to get at least 8 hours of sleep so she is ready for another day's work. However, three times a week she joins her friends at a local bar where they sit and talk and Anna will drink 3 or 4 tall beers. On these occasions she smokes cigarettes since most of her friends smoke too.

Recently Anna's mother has fallen ill. The doctor thinks Anna's mother has suffered a very mild heart attack and needs to rest. Anna has a family history of heart disease; her grandmother died of a heart attack several years ago. Anna is worried about her mother, and Anna realizes that she needs to make some lifestyle changes.

Questions

1. List all of the factors that influence Beth's and Anna's food choices.

2. Which of these factors do you think was most influential for Beth and for Anna?

3. What risk factors for chronic disease do Beth and Anna exhibit?

4. Looking at the food and lifestyle choices Beth and Anna make on a typical day, how might you advise them to modify their diet and daily routine?

5. How might fried chicken dinners and Beth's convenience freezer foods contribute to diet related disease?

6. Assume Beth's breakfast consists of 31 grams of carbohydrate, 11 grams of protein, and 8 grams of fat. Anna's breakfast has 51 grams of carbohydrate, 21 grams of protein, and 14 grams of fat. Calculate the total calories in each of their meals.

7. Clearly, there are differences in food choices between Beth, an American, and Anna, a Czech. What type of research design studies the difference between groups of people?

Case Study #2: Planning a Healthy Diet

In her book, *Nickel and Dimed: On (Not) Getting By in America,* author Barbara Ehrenreich wanted to understand how a single mother might manage to live on low wages after the welfare reform bill was passed in 1996. Ehrenreich found a minimum wage job and a place to live, and attempted to eat and survive for one month in three different cities in the United States. At one point she was desperate enough to go to a food pantry and received an emergency food package. Following are contents similar to those of the package she received:

- 21 cups of Corn Chex cereal
- 24 cups of Grape-Nuts cereal
- 2 cups of catsup
- ¼ cup butterscotch morsels
- 1 cup gumdrops
- 2 single-serving bags of jellybeans
- 2 sweet dark chocolate candy bars
- 1 dozen sugar cookies
- 6 hamburger buns
- Six 8-ounce cans of fruit punch (with 10% real juice)
- One loaf (16 slices/25 g each) of enriched Vienna bread
- 1 box (8 small) fruit roll-ups
- One loaf (18 slices) of enriched raisin bread
- 18 ounces of peanut butter
- 16 ounces of canned ham
- 1 package (6 servings/4 bars = 1 serving) of fig bars
- Two Ritz cracker packages (4 servings/12 crackers per package)
- One 5-ounce can Swanson chicken broth
- 2 ounces of a Kool-Aid-like drink mix (makes 8 cups)

Questions

1. Place the foods from the given list into the appropriate food group.

Breads, cereals, grain products	Vegetables	Fruits	Meat, poultry, fish and alternates	Milk, cheese and yogurt

2. Evaluate your groupings. Which food groups are over- or under-represented? Give the food pantry some suggestions to better help their patrons.

3. Evaluate the foods in the emergency food package using as many of the six basic diet-planning principles as apply. Nutrient contents of all foods listed are found in the food composition tables in Wadsworth's nutrition texts. They can also be found on Wadsworth's *Diet Analysis Plus* software, or online at the USDA's web site. http://www.nal.usda.gov/fnic/cgi-bin/nut_search.pl

4. Many of the foods provided give empty kcalories. Explain what is meant by "empty kcalories" and identify the foods this term applies to.

5. What is a healthier dessert than the sugar cookies or fig bars? Why? (Refer to the nutrient contents.) Nutrition data can also be found on diet analysis software packaged with the text, or online at the USDA's web site. http://www.nal.usda.gov/fnic/cgi-bin/nut_search.pl

6. Pick a food off of this list and describe the health claims the manufacturer could make. Again refer to the nutrient contents.

Case Study #3: Distressed Digestion

This is Tyler's freshman year in college and he has been earning very good grades and performing well in his part-time job. However, Tyler's eating habits have been poor and he has gained a few pounds, and his expanded waist line is making his trousers fit pretty tightly. His daily routine holds clues to his recent weight gain.

Tyler begins each morning about 9:00 a.m. when he hurriedly dresses and rushes to class without eating breakfast. He does however drink a 20-oz. cola each morning for caffeine and sugar energy. All day he sits in classes or studies in the library. About four in the afternoon Tyler takes a break for some food and he usually heads to the food court and gets a large order of chicken wings with extra spicy hot sauce. He doesn't have much time though, so he rushes through the meal, drinks another large cola, and is off to work. Tyler has a job as a waiter in a popular campus restaurant.

When the restaurant closes, Tyler heads home in his car. He usually stops to get some dinner, even though it is 11:30 p.m. He typically picks up a large pepperoni pizza that he consumes in one sitting. Sometimes he follows the pizza with handfuls of cookies that he eats while lying down on the couch and watching late night television. Lately he finds it difficult to fall asleep though. Tyler has been experiencing heartburn and acid reflux at night. His stomach makes some atrocious noises and he's been feeling bloated. Tyler has tried getting up and taking antacids, but he still suffers indigestion. Recently, a friend, who has noticed Tyler's odd stomach noises and hiccupping through class, suggested Tyler do something about his diet. At first he ignored the advice, but now his stomach pains are getting so bad he is willing to try anything.

After another painful night's sleep, Tyler wakes up early, goes to the bakery, and buys two bran muffins for breakfast, thinking this is the start of his new healthy diet. To Tyler's surprise this only makes him feel worse. He suffers horrible cramps, and later that day is forced to run to the bathroom with diarrhea. The worst part for Tyler was when he was sitting in English class and flatulence hit. Poor Tyler turned bright red and decided it was time to take the rest of the day off and go home.

When he got back to his room the phone rang. It was Tyler's grandma. He mentioned he wasn't feeling too well, and Grandma, who lived nearby, came over with chicken noodle soup made with carrots and celery and lots of broth. The two had a leisurely meal together, sitting and talking at the table for hours. Tyler ate one small bowl of soup and placed the leftovers in the refrigerator for later. Tyler then got a good night sleep, and woke up the next day feeling much better. Once again, his fast-paced lifestyle plunged him back into previous eating patterns. On the way to his first morning class Tyler was drinking a 20-oz. cola.

Questions

1. Name the symptoms Tyler was experiencing initially and describe the probable causes of his ailments.

2. He tries taking antacids. Is this a good idea? Why or why not?

3. After adding bran muffins to his diet, Tyler only felt worse. Why is this?

4. After Tyler's grandma visited and they ate some chicken soup together, Tyler felt much better. Why might this be?

5. Tyler experiences the unpleasant and embarrassing consequence of poor nutrition while in English class. This only adds to the already high stress in Tyler's life. What was the cause of this and how can Tyler cure this problem for the future?

6. What advice would you give Tyler about his eating habits?

7. What do you think the long-term consequences of Tyler's eating habits might be?

Case Study #4: Simple Sugar and Complex Carbohydrate

Clara is an African American college student. She is 5'4" and weighs 170 lbs. Clara's entire family is overweight and her father was diagnosed with diabetes last month. Her uncle has been suffering from diabetes since he was eight years old. Clara fears that if she does not change her diet she may develop diabetes as well.

On an average day Clara eats two jelly doughnuts for breakfast, chicken nuggets with sweet and sour sauce and a large soda for lunch, and goes to the drive-through to grab a super-sized burger, fries, and a large cola. She usually snacks on jellybeans during the day and is always sipping on a cola.

Clara is concerned about her diet and decides to ask her friend Mary, nutrition major, for advice. Mary gives Clara some suggestions. First she tells Clara that, while her diet is very high fat, it is also very high in carbohydrate. "I should try that diet advertised on TV. If I cut all carbohydrate I can lose 50 pounds this month!" Clara announces. Mary rolls her eyes. Mary explains that complex carbohydrates are good, but too much simple sugar is not. Mary tells Clara that she should snack on fresh fruits instead of jellybeans, and eat more starch and fiber. Plain baked potatoes (instead of fries) and fortified cereals (instead of high-fat doughnuts) would be a lot healthier. Small changes like these could improve her diet dramatically.

"Try not drinking so much soda," Mary suggests. "Bottled water or a glass of milk would be much more beneficial."

"I can't drink milk!" Clara says. "Last time I had a glass of milk my stomach hurt for hours. I think I'm lactose intolerant since it runs in my family."

Clara and Mary discuss cow's milk alternatives and Clara thanks Mary for the advice and goes home to plan her new healthy diet.

Questions

1. Diabetes clearly runs in Clara's family. Differentiate between her uncle's and her father's diabetes. How do Type 1 and Type 2 diabetes differ?

2. What form of diabetes is Clara at risk of? What is putting her at risk?

3. Clara thinks she can't drink milk because she is lactose intolerant. How might Clara include milk and milk products in her diet?

4. If Clara does not want to give up soda entirely, but still consume less sugar, she could drink diet sodas. What are the pros and cons of this switch?

5. Clara discovers that a lot of her calories come from carbohydrate. She initially decided to eliminate all carbohydrate from her diet, but Mary tells her that this is a poor decision. Why?

6. Although there may be the same amount of sugar in fruit as some candies, why is fruit a much better choice for Clara?

7. Mary suggests eating more fiber. How would this be beneficial in Clara's case?

Case Study #5: Not Too Much and Not Too Little: Understanding Fats in Foods

Jenny and Travis are four months into their freshman year of college. The two have been friends since grade school, and find a great deal of comfort in being able to get together for dinner.

Travis loves food, but rarely tries new things. At first, the unfamiliarity of the school dining hall was scary, but once he discovered that he could get a steak and cheese sub seven days a week he was content. For Jenny the adjustment was much more difficult. Jenny has been a semi-vegetarian for many years, and the vegetarian entrees cooked at the dorm are not particularly inspired. Although she eats fish, the cafeteria rarely serves fish that isn't breaded and deep-fried. Typically Jenny sticks to salads with low-fat dressing, low-fat yogurt, or a bowl of cereal with low-fat milk for dinner.

"I'm starving!" she told Travis one night at dinner. "There is never anything I want to eat in the dorm and I'm losing a lot of weight. This winter the colder temperatures are bothering me more than usual and I have been so tired. I can't wait for this weekend. I'm going home to have a home cooked meal." Jenny couldn't help but smile as Travis's eyes lit up at the mention of a home cooked dinner. "You want to come?" she asked, knowing the answer would be a resounding, "Yes."

When they arrived the house was filled with the wonderful aromas of home cooking. There were hot rolls, salad with olive oil and vinegar, grilled salmon, and fresh chocolate chip cookies for dessert. They sat down to dinner, and after just one bite Travis decided this was the best meal he had ever had. He didn't hesitate to ask for a second helping of everything. Jenny ate a big meal as well. As she spread some margarine on her roll, she couldn't help but smile seeing her best friend enjoying his dinner.

Then came dessert. Jenny's parents served the chocolate chip cookies with some low-fat frozen yogurt, both of which were delicious. "Now, I made the cookies with real butter," Jenny's father reminded his wife, who had been cautiously watching her cholesterol. She was stuffed after a big meal anyway, so she skipped the cookies. Still, she enjoyed watching the kids devour their desserts. "I can't thank you enough," Travis remarked as they were leaving to go back to school. "Any time you want to come over for dinner, Travis, you are always welcome," Jenny's mother replied, and Travis fully intended to take her up on the offer.

Questions

1. Travis eats steak and cheese subs every day of the week. Explain the types of fats, as well as the properties of each type, that are found in these subs.

2. Jenny eats low-fat salads, yogurt, or cereal at every meal, and she is feeling tired and more sensitive to cold temperatures. Explain what dietary factors may have caused some of the symptoms she experienced.

3. Jenny and Travis ate salmon for dinner. What types of fat are found in this food? How does the fat found in salmon differ from that found in steak and cheese subs?

4. Olive oil and vinegar are used to flavor the salad. Explain some of the benefits associated with olive oil.

5. Jenny spread margarine on her roll, although the cookies were baked with butter. Compare and contrast these two fat sources.

6. Jenny's mother has been watching her cholesterol. She was warned not to eat the cookies because they were made with butter, which contains cholesterol. However, chocolate chip cookies contain other sources of cholesterol. What are they? (Ingredients: Flour, sugar, butter, eggs, milk, vanilla, baking powder, milk chocolate chips.)

Case Study #6: Vegetarian Diets

Dawn is a 22-year-old college senior. She is a vegetarian who believes strongly in animal rights. So, over the last three and a half years she has eliminated all meat and animal products from her diet. Dawn has also lost a little more than 15 lbs. Lately she feels that she has lost too much weight. Dawn is 5'3" and weighs 105 lbs. Her diet includes alfalfa sprouts, legumes of various types, and tofu that has not been fortified. She especially loves black bean soup, peanut butter sandwiches with multi-grain bread and several soy products. Dawn occasionally drinks soymilk. The only vegetable oils that she uses are safflower oil or canola oil. She does not take vitamin or mineral supplements, such as calcium, because she believes they are not necessary. Her theory is that she can obtain all her vitamins and minerals from the food she eats. Dawn drinks no alcoholic beverages and does not smoke. She believes it is best to eat only "natural" foods. She eats no bleached flours and eats whole grain cereals and bread. However, she makes sure to eat plenty of fresh fruits and vegetables at every meal.

Despite her health-conscious measures, Dawn catches colds easily and has a harder time fighting them off. Dawn has never been very active, but lately she has lacked energy. Dawn wonders if she should look into her diet. She decides it is time to investigate the facts about vegetarian diets in more detail, with the hope that she can feel better through more informed vegetarian eating.

Questions

1. What kind of vegetarian is Dawn? Name and describe some other types.

2. Examine Dawn's diet and give her some advice on how to eat a healthier diet without giving up her strong beliefs about animals.

3. Dawn eats significant quantities of sprouts, peanut butter, and beans. What are some of the advantages and disadvantages of these protein sources?

4. Name the sources of calcium in her diet.

5. Dawn feels that vitamin/mineral supplements are unnecessary; do you agree?

6. Aside from vitamin/mineral supplements, some people take protein and amino acid supplements. Explain some of the dangers associated with this.

7. By eating better, Dawn hopes to gain a few pounds. Calculate the amount of protein Dawn should eat per day, if she is 5′3″ tall and hoping to weigh 121 lbs.

8. Design two one-day meal plans for a 5′3″, 121-pound female who is very lightly active. Make one diet meat based and the other a vegan diet appropriate for Dawn. Compare the two in terms of calcium, iron, protein, and total calories. Use a food composition table from a textbook, *Diet Analysis Plus* software, or the USDA's nutrient database at http://www.nal.usda.gov/fnic/cgi-bin/nut_search.pl .

Case Study #7: Feasting and Fasting

Kathleen is a 5'6", 130-pound, 20-year-old college junior. Over the last few years she has gained then lost ten to fifteen pounds several times. Recently she has been trying hard to keep her weight down. Like many college students, Kathleen goes out every weekend and tends to overindulge. During the week she lives the life of a serious student, eating very little and getting 7-8 hours of sleep every night. By Friday she is ready to have some fun. She usually goes out with friends to a nearby burger or taco shop and then finds a party. At the party Kathleen will have four or five drinks and munch on chips or pretzels. By 3 a.m., when she and her friends are ready to call it a night, they usually are hungry again. This means finding pizza, waffles, or any other food they can find at that hour of the morning. After a late night out, Saturdays are spent catching up on some sorely needed sleep. By Saturday night she is ready to go out and do it all over again. Sundays are usually spent relaxing at a hearty Sunday brunch, watching movies with her friends accompanied by a giant bowl of popcorn, and finishing homework that's due Monday morning.

Kathleen realizes that her weekend binges may cause her to gain weight, so she cuts way down on calories Monday through Thursday. Kathleen has been trying to stay active and build up some muscle mass by running. Recently, her strict dieting is making this more difficult. During the week Kathleen eats so little that she often feels weak or lightheaded, especially following her long distance runs. Although she is running a great deal, her muscles are not getting as large as she had hoped. On Mondays and Tuesdays her diet is very hard to stick to. Kathleen is always starving. However, by the end of the week she no longer feels so hungry. Kathleen also notices that she has a much harder time paying attention, is sensitive to the cold temperatures, and finds it all too easy to catch a cold or flu.

For a while she was able to overcome this and still go out on the weekends, but she is starting to feel that it is not worth being so miserable during the week, no matter how much fun her weekends may be. Kathleen knows she is putting her body through a lot of stress, but needs someone to explain what is happening as she experiences the consequences of a weekly roller coaster ride from feasting to fasting.

Questions

1. After Sunday brunch Kathleen's digestive system is filled with nutrients. The liver receives these nutrients first. Describe what the liver does with the carbohydrates, fats, and proteins in Kathleen's Sunday brunch.

2. On the weekends Kathleen drinks substantial amounts of alcohol. Explain alcohol's effects on the liver, both short term and long term.

3. During periods of fasting, how are metabolic fuels used differently compared to their use with a healthy, consistent diet? Describe some of the negative effects associated with this.

4. During periods of feasting, how are metabolic fuels used differently compared to their use with a healthy, consistent diet?

5. Kathleen has a lot of warning signs that she is not eating well. Name her symptoms and explain why she feels this way.

6. Kathleen's weekend diet consists of many more calories than during the week. Discuss the body's reaction to excess carbohydrate, protein, and fat.

7. What are the long-term consequences of Kathleen's eating pattern and alcohol consumption on her vitamin status?

Case Study #8: Satiation and Appetite

Gina has been head chef at a local Italian restaurant for the last 15 years. She is from a large Italian family that loves not only cooking, but also eating. Although Gina has grown up around large amounts of food and hearty appetites, she has learned to control her weight and eat moderately. However, Gina's husband Anthony has difficulty moderating his eating habits. They have been married for 3 years and in that time Anthony has put on 40 pounds, most of which went straight to his belly. As a software analyst, Anthony gets very little exercise. To make matters worse, Anthony often has large, rich meals with his in-laws. Last week they celebrated Anthony's 55th birthday with pasta in cream sauce, breaded veal, and lots of garlic bread drizzled with olive oil. They washed the meal down with several glasses of wine, and devoured cannolis for dessert. On this birthday, Anthony realized that he really needed to change his eating habits. He is now 55, about the same age his father and grandfather both died of heart attacks, and in addition he has recently learned that both of his brothers have high cholesterol. Anthony has never had his blood cholesterol checked, but wouldn't be surprised if his levels are high too.

Anthony went to his family doctor for a physical. Anthony weighed in at 215 pounds, substantially overweight for his 5'11" frame. The doctor asked Anthony to describe his daily routine and eating habits.

Anthony reported that he sleeps late and needs to rush to work. He rarely eats breakfast although he drinks coffee with cream and sugar. Around noon everyday Anthony's coworkers go to the sub shop across the street. Anthony never joins them because he says the subs are not "real" food. Anthony does not eat due to boredom or anxiety; he eats because he loves food. This is why he brings a freshly made sandwich from home. His sandwiches are usually made of the leftover chicken parmigiana or veal cutlets from the night before. Anthony brings very small sandwiches to merely tide him over until the evening's feast.

When Anthony gets home he likes to relax by helping to prepare the evening meal. Smelling the delightful aromas and looking at the mouth-watering ingredients only increase Anthony's appetite. Dinner is the meal he lives for. When he sits at the table in the evening, he leaves the day behind and has nothing to do but enjoy a good meal and go to bed.

His typical supper consists of a tossed salad with an olive oil and vinegar dressing followed by pasta, with tomato sauce and grated cheese, or cheese sauces. Fresh mozzarella with sliced tomatoes is usually served as well. Anthony likes Italian sausage or veal with his pasta. Anthony fills his plate with pasta, has a second helping of sausage, and can't resist a bowl of gelato, which is always found in their freezer, for dessert. Anthony knows he needs to change his eating habits, but refuses to deny himself the foods he loves, so he asks the doctor for some advice.

Questions

1. The doctor begins by determining Anthony's BMI. Calculate Anthony's BMI, recalling that he is 5'11" and 215 pounds.

2. What would a healthy BMI for Anthony be? What is the corresponding weight range?

3. Knowing that Anthony does not engage in much physical activity, estimate his daily energy output.

4. We know a little about Anthony's family history; why is this particularly alarming?

5. If Anthony does not change his diet soon, what health risks, in addition to cardiovascular problems, does he need to think about?

6. Anthony is easily able to overcome hunger signals during the day. What then does influence his food intake?

7. Although satiation sets in about 20 minutes into a meal, Anthony usually does not respond to this. What changes to his diet could influence the satiation he feels?

8. Anthony was much thinner just three years ago. Although much of his weight gain has to do with his new eating habits after marriage, other factors contribute to it. What are some of these factors?

9. What general advice should the doctor give Anthony regarding his diet?

Case Study #9: Diet Strategies for Overweight and Obese Individuals

Mickey and Jim are brothers. Their parents, along with most other members of their family, are overweight and have hypertension. Mickey is the youngest of the family and at age 24 is overweight. He is 5'10" tall and weighs 188 lbs. Mickey is active, although not as active as he should be. He works at an advertising agency and is on the company softball team. They have been practicing every Saturday for the last three months. From 11 a.m. – 2 p.m. Mickey is running bases, hitting balls, and playing the outfield. Afterwards, he and his friends go to the local pub for a few beers and a bite to eat. This "bite" is usually a burger and fries.

Mickey's brother Jim is not active at all and is 6′ tall and weighs 338 pounds. Jim is a middle school history teacher. Most of his time is spent behind a desk or sitting at the table grading papers. When he needs a break he usually goes to the fridge for a snack. Having grown up in the same household, Mickey and Jim have similar food preferences. Both love old-fashioned eating: bacon and eggs, ham and cheese, steak and potatoes. As kids, neither brother wanted to eat his vegetables, although they do eat some vegetables now. One habit they fully enjoy and never grew out of is snacking on milk and cookies. To this day they both have milk and cookies before bed. They were pudgy children, although they did not seem to eat much differently from the other kids in the neighborhood; they just gained more weight than the other kids on the block.

Mickey loves food just as much as his brother, but he knows his weight is not healthy and that his brother's is even worse. So, he asks Jim to go on a diet with him. At first Jim seemed reluctant, but now he has turned Mickey's weight loss challenge into a brotherly competition. Both are determined to lose at least 45 pounds in the next 10 weeks.

Mickey decides that he needs to exercise more to lose weight. In addition to his Saturday softball practice, he plans to jog for 15 minutes twice a week. This does not produce results quickly enough, so Mickey tries using diet pills, but only "natural" ones. Mickey tries St. John's wort but has tremendous difficulty sleeping at night. He also takes dieter's tea but can't deal with the nausea and cramping. Finally he decides to go to his doctor and get prescription drugs. The doctor tells him that sibutramine and orlistat are the most common drugs on the market, but he feels orlistat is the better choice for Mickey.

Jim also goes to his doctor, but does so before attempting to change his activity level or trying natural diet products. He asks his physician if he would be a good candidate to have his stomach reduced in size by surgical stapling. Jim schedules an appointment for a consultation with the surgeon, but never sees it through. The brothers have a long talk and decide they should change their diets and eating habits to obtain healthier, more long-term goals.

Questions

1. Losing 45 pounds in 10 weeks is not realistic. A more appropriate goal would be to lose 10% of their body weight in 6 months (24 weeks). How much weight should Jim and Mickey aim to lose?

2. As children Mickey and Jim did not eat any differently from other children but gained more weight. Why is this?

3. While growing, Mickey and Jim were overweight. How could this affect their current weights?

4. At first, both brothers set unrealistic goals, something that often leads to yo-yo dieting. Explain why this is detrimental to one's health.

5. Mickey initially tried St. John's wort and dieter's tea but experiences unpleasant effects. What are these effects and why did they occur? What about the St. John's wort could be especially dangerous?

6. Mickey is substantially overweight, to the point that his doctor was willing to prescribe medication. Why did the doctor recommend orlistat but not sibutramine?

7. Jim did not want to waste his time with conventional weight loss methods and immediately considered surgery. Calculate both Jim's and Mickey's BMIs. Based on these numbers why is Jim a more suitable candidate for this procedure than Mickey?

8. Do you think surgery is a wise choice for either of the brothers? Why or why not?

9. What behavioral changes do both Jim and Mickey need to make in order to control their weight?

Case Study #10: Absorption of Water-Soluble Vitamins

Brian is a 21-year-old college student who weighs 170 pounds and is 5 feet 10 inches tall. He lives with several roommates in a house near his campus. Brian leads a moderately active lifestyle by playing basketball and occasionally lifting weights at the campus gym. He enjoys eating pizza or burgers and drinking beers with his friends. However, Brian thinks that the pizza and burgers are not healthy foods, so occasionally he changes dietary gears. For a day or two Brian nearly fasts. However, on these fasting days he still allows himself to drink beer, and once a day he has a small bowl of cooked carrots and peas.

Generally Brian hasn't been feeling his best: he has trouble sleeping, especially when he does his "fast," and notices his gums bleed after brushing his teeth. To figure out what is causing these symptoms he looks at his food intake on a particular, and typical, Saturday. It is as follows:

Breakfast: 10:00 a.m. (at his house with friends)
Two frosted strawberry Pop Tarts
Super 20 ounce Gatorade

Lunch: 1:30 p.m. (at a nearby McDonald's)
Big Mac
Supreme Super Size fries
Super Size Coke
Apple pie

Snack: 5:00 p.m. (at his house with friends)
Two cans Bud Light beer

Dinner: 7:00 p.m. (at home)
Three slices of Pizza Hut Supreme pan pizza
Eight cans Bud Light beer

Snack: 1:30 a.m. (at home)
One slice Pizza Hut Supreme pan pizza

Here is an analysis of his diet:

Name:	Brian				
Analysis of:	1 day (plus comparison to RDA)				
Nutrient	**Amount**	**% RDA**	**Nutrient**	**Amount**	**% RDA**
Calories	4446	153	Vitamin C	51.5 mg	86
Protein	114 g	197	Vitamin D	1.06 mcg	11
Dietary fiber	18.9 g	---	Vitamin E	21 mg	210
Vitamin A total	424 RE	42	Calcium	1600 mg	133
Thiamin—B_1	3.18 mg	212	Iron	24.8 mg	248
Riboflavin—B_2	3.29 mg	194	Magnesium	361 mg	103
Niacin—B_3	45 mg	237	Phosphorus	1670 mg	139
Vitamin B_6	2.41 mg	121	Zinc	12.9 mg	86
Vitamin B_{12}	5.2 mcg	260	Alcohol	113 g	---
Folacin	235 mcg	118			

Exchanges			
Bread	20.8	Vegetable	2
Meat	6	Milk	2
Fruit	1.6	Fat	36

Calorie Breakdown	% kcal
Protein	10
Carbohydrate	45
Fat	27
Alcohol	18

Questions

1. Which water-soluble vitamins does Brian have too much or too little of?

2. Would you recommend that Brian take supplements for the vitamins he lacks in his diet? Why or why not?

3. Why does Brian have trouble sleeping, especially when he restricts his diet?

4. What can he do to stop his gums from bleeding so easily?

5. How might Brian's alcohol intake be affecting his nutrition status?

Case Study #11: Environments and Susceptibility to Vitamin D Deficiency

Sharon and Janette are sisters. While in high school they each decided to adopt vegan diets. Sharon and Janette plan their diets carefully and do not take a nutritional supplement, preferring to get their nutrients from food. Neither drinks fortified juices or soymilks, again, preferring to get nutrients "naturally."

Sharon recently turned 30. She has two young children and is pregnant with her third child. Her husband, Russell, is in the military, and a few years ago the family moved to a base in Alaska. Sharon does not like the cold weather and long, dark winters. She has been working in a local factory and does a lot of physical labor. Recently, Sharon has noticed a lot of pain in her lower back and legs. It is far too early to be a result of her pregnancy, and Sharon is becoming concerned.

Janette, who moved to southern California after college, is the younger of the two sisters. She was married in California, started a family, and is embarking on a new career. Janette works the night shift on a newspaper as a copier. Her job is to assure that the paper comes off the press and is ready to be delivered first thing in the morning. Janette's husband is able to take care of their son while Janette sleeps during the daytime. They are discussing having another child, but Janette does not think that her inverted day-night schedule will be suitable for her growing family. She also fears for her own health if she has another child. Janette has had several bone fractures in the last few years, and recently has had a lot of pain in her pelvis. Although she wants to have another child, she needs to be examined by her doctor first.

Sharon and her sister talk regularly on the telephone. They look forward to comparing lives as well as ailments. Both of them know that something needs to be done to better understand their health problems.

Questions

1. What might be the cause of Sharon's leg and lower back pain?

2. What might be the cause of Janette's pelvic pain and easily fractured bones?

3. Is there anything Sharon could do to help her ailments?

4. Is there anything Janette could do to help her ailments?

5. Each sister is having or is considering having another child. Why could pregnancy and repeated periods of lactation have adverse consequences on their health?

6. Are the children of these sisters at risk of similar health problems? If so, name the most probable diet-related disease they will develop.

7. Which child is at greater risk and why? What should these mothers do to prevent their children from developing dietary health problems?

Case Study #12: Hidden Sodium

John is a 35-year-old advertising executive. He is fairly active, has a BMI of 23, and is in generally good health. However, at his yearly physical his doctor discovers that John has high blood pressure. John assures his doctor that he eats well, and can't understand why the numbers are so high. John's doctor tells him that he probably is salt sensitive, and needs to cut back on the sodium in his diet. When John returns home he immediately puts the saltshaker in the back of the cupboard, so he is not tempted to add salt to his foods. He also replaces his usual snack of salted peanuts with dry cereal, his favorite being cornflakes. He has heard about packaged foods being high in sodium, so he swears off his usual frozen dinners and takes the time to prepare some of his favorite meals. John eats Mexican rice and beans, or a piece of smoked halibut with mashed potatoes to go with it. Since lunch is usually at the office, John decides he needs to start bringing lunch instead of ordering Chinese food with his coworkers. John goes out to the store and buys a package of bologna and some American cheese to make sandwiches. He enjoys a pudding cup for dessert. He is always very thirsty so he brings both a bottle of water and orange juice to sip on, but rarely milk. After a month on this new diet John returns to his doctor, assured that his blood pressure has gone down. To his surprise, the numbers have not budged. "I don't understand! I don't put salt on anything," John tells his doctor.

Questions

1. Explain how sodium is the contributing factor to John's hypertension.

2. The first step John takes in lowering his sodium intake is hiding the saltshaker. Do you think this was very effective? Why or why not?

3. John replaces his snack of salted peanuts with corn flakes. Using a food composition table, look up the sodium content in both and explain why this was or was not a good choice.

4. John is aware enough to know that packaged foods and Chinese food can both be extremely high in sodium. What makes these foods high in sodium even if they are not particularly salty in taste?

5. He replaces frozen dinners with Mexican rice and beans or smoked halibut, two seemingly healthy choices. Compare their sodium contents to the RDI.

6. The sodium in John's diet makes him thirsty. Explain why this is.

7. John generally drinks water or juice to quench his thirst, but rarely milk. Although he does have some dairy from pudding or cheese, he does not ingest much. How could this, in conjunction with his high sodium levels, have adverse effects on his health?

8. Design a diet for John that would lower his sodium intake, improving his blood pressure.

Case Study #13: A Little Is Good, but More May Not Be Better—Iron Deficiency and Toxicity

Laura is a twenty-one-year-old communications major at a small college just outside of Duluth, Minnesota. She is an average student in the classroom, but excels in intramural track and field competitions. She is about to start her senior year, and Laura wants to maintain the school track and field records, which she has earned in the past few years. Because she wants to be in top physical condition when the school year begins, Laura changes her diet so that she now only eats low-fat, highly nutritious foods. She has become a vegetarian. Although she still eats eggs and dairy products, she now chooses soy patties over hamburgers. Laura also begins running every morning before school.

She was in the best shape of her life the first few months of the school year, winning all of her events. Recently, however, Laura has not been feeling too well. She has noticed that it is much harder to gather the energy for class every day. She has a hard time "getting up" to start new things. At first, she noticed it only once a month, while she was menstruating, but now she feels tired and weak more often. She gets particularly tired after running sprints. At the beginning of the year she set the pace for her team, but now she finds it difficult to keep up.

At first, she thought she was just tired and stressed from such an active lifestyle, but now she is noticing other symptoms, causing her to be a bit more worried. She is now experiencing shin splints (persistent pain in the shins). Lately she has looked pale. She gets frequent headaches and finds herself much more irritable. This may be because she has been having trouble sleeping lately. This winter has also been particularly cold and Laura can't seem to warm up. Laura's symptoms were not severe enough for her to worry, so she put off seeing the doctor. However, Laura now tends to get more bruises than she used to. And, when she gets cut it takes much longer to heal. When she does get a cut it often becomes infected. Laura decides it is finally time to see her physician.

While at the doctor she has blood drawn and analyzed. Her results are sent in the mail later that week. She receives the following information:

Laura's Test Results

Hemoglobin	110 g/L
Hematocrit	.30
Mean Corpuscular Volume	70 FL
Mean Corpuscular Hemoglobin	25 pg
Serrum Ferritin	10 ug/L
Erythrocyte Protoporphyrin	1.36 umol/L RBC

Lower Limits of the 95% Reference Range for a 20—44-Year-Old Female

Hemoglobin	120 g/L
Hematocrit	0.35
Mean Corpuscular Volume	80 fL
Mean Corpuscular Hemoglobin	27 pg

Recommended Cutoff Value for Confirmatory Test for Iron Deficiency for Someone 15 Years or Older

Serrum Ferritin	Less than 12 ug/L
Erythrocyte Protoporphyrin	Greater than 1.24 umol/L RBC

*P.R. Dallman. 1987 Iron deficiency and related nutritional anemias. In: Hematology of Infancy and Childhood 3rd ed. (D.G. Nathan and F.A. Oski eds.) pp. 274-296, Saunders, Philadelphia.

by Thomas W. Castonguay

Laura can tell that her iron is low, so before discussing it with her doctor she immediately goes to the store and buys iron supplements. She takes a 60 mg supplement before every meal and again before going to bed. Laura makes sure to swallow the supplement with a glass of orange juice because she knows Vitamin C enhances iron absorption. She also tries to eat more iron-rich foods, without changing her vegetarian diet. She loves spinach, and, having read that it is high in iron, finds a way to incorporate it in most every meal.

Laura quickly begins to feel better, but this doesn't last for long. After one month her symptoms return and additional symptoms develop. Laura has noticed that she was constipated when she first began increasing her iron, but lately she has had nausea and diarrhea. She feels just as weak and lethargic as before, and her occasional cuts and scrapes continue to be easily infected. Laura begins to wonder: if it's not an iron deficiency, then what is wrong with her? She knows something has to change, but needs help figuring out what to do.

Questions

1. Why might Laura's symptoms have returned, even after changing her iron consumption?

2. Laura experiences many symptoms that eventually bring her to the doctor. Which of these are indicative of iron deficiency?

3. At first Laura only experiences symptoms during menstruation. Explain why this is.

4. Interpret the results of Laura's bloodwork.

5. When Laura discovers that her iron is low she immediately makes dietary changes. What misconceptions does she have about iron intake and absorption?

6. What general advice can you give Laura to improve her health?

Case Study #14: Can Physical Fitness Come in a Bottle?

Ryan is a 21-year-old college student who is very interested in bodybuilding. His ambition is to compete in bodybuilding on the local and then state levels, and perhaps one day enter competition on the national level. He is 6'1" and weighs 212 lbs. Because of his lean body mass, he does not look overweight. He works out every day for two to three hours, working different body parts each time. His goal is to gain 25 pounds. To do so, he is taking dietary supplements, including protein, vitamin E, and creatine. The gym he attends has high-quality equipment and support services. Among the gym's services are cardiovascular and respiratory evaluation, fat analysis, and dietary instruction. He considers speaking with the dietitian, but her main job is working with people who are overweight, so Ryan decides not to seek her advice.

What Ryan knows about nutrition comes from various bodybuilding magazines. He has put himself on a high-protein, low-carbohydrate, low-fat diet. He obtained a calorie counter book from a newsstand and uses it to determine the amounts of carbohydrate, fat and protein he eats. He has discovered that he takes in 10 percent or less of his total calories from fat. Ryan is eating 150 g of protein per day. He understands that the best form of protein is egg, so his 150 g of protein includes four raw eggs per day. In addition, he eats 3 g of protein powder before working out, 3 g after working out, 3 g in the morning, and 3 g in the evening. The protein powder is a combination of amino acids he adds to skim milk. He is also taking a high-stress multiple vitamin and mineral tablet every day. The tablet contains 200 percent of his requirement for vitamins A and D. The vitamin A is in the acetate form. In some of his magazines he has read that vitamin E prevents oxidative stress caused by physical activity, so he takes a Vitamin E supplement each day. He has also read that Vitamin C is necessary for protein synthesis, so he takes 2 g of Vitamin C a day. Both are in addition to the multiple vitamin he is taking.

One day, Ryan decided to use the gym's service of bioelectrical impedance testing to measure his body fat. To his surprise Ryan discovered that he has 10 percent body fat. He was disappointed with this and wanted to get his body fat down to 7 percent. To do so he decides to begin a cardiovascular fitness program. He begins jogging each morning, but quickly feels light-headed and tired. He doesn't understand why, because he is able to do weight training exercises for hours without any adverse effects. Ryan decides to ask one of the trainers at the gym about this. The trainer suggests it would be a good idea to talk with the gym's dietitian.

Questions

1. What is your opinion of Ryan's current diet? Why?

2. Ryan is on a high-protein, low-carbohydrate, low-fat diet. Does this help or hurt his goal of gaining 25 pounds? Explain.

3. Ryan is taking a multi-vitamin as well as extra supplements and a variety of foods. Refer to the information on water- and fat-soluble vitamins in your textbook. What adverse effects can these extra supplements have?

4. Ryan is lifting weights for several hours a day. What type of exercise is this? Explain what this has to do with a fuel source and protein use.

5. Why does Ryan's diet give him energy for weight lifting but not for jogging?

6. Do you think Ryan's goals of gaining 25 pounds and reducing his body fat to 7% are realistic and healthy? Why?

7. Describe a better diet and exercise program for Ryan.

Case Study #15: Nutrient Needs of Women and Infants

Cameron and Bradley are 25-year-old, budding Hollywood stars. They live in a large house where they host parties almost every week. Recently, Cameron decided that she is ready to become a mother. She and Brad want only the best for their child and are willing to change their lifestyle to help ensure the arrival of a healthy baby. Cameron goes to her doctor to get some advice. First, the doctor weighs her and finds that she is 5'4" and only 100 lbs. He asks about her diet and finds that she lives mostly off of coffee and cigarettes during the day and wine and liquor every night. Most of her caloric intake from food is eaten as hors d'oeuvres. She is physically inactive, never getting any exercise. The doctor can't help but be critical, and tells Cameron that if she is serious about ensuring the health of her baby she will need to change her lifestyle drastically before even attempting to become pregnant.

When Cameron returns home from the doctor she and Bradley discuss the matter. Having a family is very important to them, so they agree make changes right away. First, they both quit smoking. They also stop throwing parties and drinking alcohol. Cameron focuses on improving her diet. She eats three balanced meals per day, stops drinking coffee, and manages to put on a few pounds, bringing her to a healthy body weight (an additional 10-15 pounds). She is particularly deliberate in making sure that she eats several servings of fruits and vegetables every day.

A few months later she goes back to her doctor. He is impressed with her dedication and tells her that she is now ready to conceive a child who is much more likely to be born healthy. Only one month later a pregnancy test turns up positive. Her doctor confirms that she is pregnant and prescribes prenatal vitamins. He strongly urges her to continue her healthy habits.

Over the next few months Cameron goes through the typical experiences of being pregnant: nausea, constipation. But, on the positive side, she has shown a healthy weight gain. Food cravings emerge, and Brad is regularly sent to the store for pickles, ice cream, apple juice, and potato chips.

As her nine months come to an end, Cameron begins to consider whether or not she will breastfeed her baby. She is not sure if this is a good idea, so she calls her sister Annie in Utah. Annie is very different from Cameron. She lives a relatively simple life, and has two healthy children of her own. Cameron asks Annie if she breastfed her children. Annie tells her that she did not breastfeed, but if Cameron wants to there are certain things she must do first. She tells Cameron that she must take vitamin and mineral supplements, particularly vitamins D, K, and calcium. Cameron tells her sister that the doctor has put her on prenatal vitamins, but Annie insists that her requirements will change once she is breastfeeding, so Cameron will have to adjust her diet.

Questions

1. If Cameron had not changed her habits before conceiving a child, what are some of the detrimental effects that could have occurred?

2. The doctor prescribed a prenatal vitamin. What does this include and why is it important?

3. What are some of the likely causes of Cameron's nausea, constipation, and strange food cravings?

4. She gains a healthy but substantial amount of weight. About how much would this be and why is it so important?

5. Do you think Cameron should breastfeed her child? Why or why not?

6. Annie has several misconceptions about the needs of a lactating woman. What are they? Give Cameron appropriate advice.

7. Design a healthy meal plan for Cameron leading up to conception, during pregnancy, and after the birth.

Case Study #16: Food Choices Differ Among Age Groups

Jenna Swanson and her husband Alex live in Oberlin, Ohio with their three children, fourteen-year-old son Kevin, four-year-old daughter Alissa, and ten-month-old daughter Amanda.

Alex is a college professor and professional musician and works very long hours. He is never home to eat dinner with the kids. Jenna is overwhelmed with caring for three children on her own and thinks that it might save her some time and effort if at each meal she served the same foods to all three children.

On Jenna's first effort at serving one meal, she plans a menu of hotdogs, grapes, and apple juice. Jenna sits all three children in front of the television, a place where they spend a lot of time. She instructs Kevin, the 14-year-old, to keep an eye on his younger sisters while she cooks.

Kevin, who is starving after this afternoon's track practice, sneaks a package of peanut butter cookies, and has a before-dinner snack. Four-year-old Alissa is all too eager to help him eat the cookies. A few minutes later Jenna calls them to the table for dinner. She has prepared hotdogs with toasted white rolls and barbeque sauce, grapes, and a glass of apple juice for each.

They sit down at the table, but Alissa doesn't eat a thing. She has been eating cookies and is simply not hungry. Kevin, on the other hand, cleans his plate in minutes. He takes about two sips of the apple juice, and then goes to the fridge for some soda.

Baby Amanda is still learning to hold her cup and spoon and Jenna helps her eat. Her hotdog is cut into small pieces but she does not like the hot dog pieces with barbeque sauce on them or the bread. Amanda does, however, enjoy the grapes and apple juice. Jenna gives her a bottle of juice after dinner, as she puts her down to bed.

Now that baby Amanda is asleep, Jenna readies Alissa for bed. In doing so, she notices that Alissa is scratching all over. Hives have developed on her back and she is complaining of a headache. In a panic, Jenna calls the doctor to find out what is wrong. The doctor calls back and calmly tells her that Alissa is probably having an allergic reaction, but it doesn't sound too severe. He tells her to use cortisone on the hives and keep an eye on her to make sure the symptoms don't get worse. Later that week he wants to conduct allergy tests to find the cause of her reaction.

Alissa's reaction subsides and she is eventually able to go to sleep. By this time Alex is home from work. He and Jenna fix themselves a late dinner of pasta, with meat sauce, dinner rolls, and salad. Kevin, who is still hungry, joins them. He eats a large bowl of pasta without the sauce and several rolls, but skips the salad. Kevin then grabs another soda from the fridge, and watches some television before going to bed.

Questions

1. Evaluate the dinner that Jenna prepared for the children. What are the pros and cons of these foods for all three of her children?

2. Why is Kevin always hungry?

3. Was this an appropriate dinner for baby Amanda? Why or why not?

4. What most likely gave Alissa an allergic reaction?

5. What important foods does Kevin lack in his diet?

6. What is the problem with giving Amanda a bottle at bedtime?

7. Alex is becoming concerned that Alissa is gaining weight. Although she is only four and he does not want to restrict her diet, he fears obesity and heart disease later in her life. Is this a valid concern? Why or why not?

8. What would more appropriate food choices be for all three children?

Case Study #17: Drug and Nutrient Interactions in the Elderly

Grandma is an amazing woman. Up until recently she was cooking, cleaning, exercising, and staying very active in her community. Unfortunately, at age 85 Grandma broke her hip. Her life suddenly became very different. She had become accustomed to her independence, but now is no longer able to get through the day without assistance. As a result, her family persuaded her to move from her home in Florida to Seattle, Washington, where she could be close to her closest relatives. Grandma was not thrilled about the move but knew it would be the best thing for her.

Once Grandma arrived, her children observed some unsettling behaviors. They saw that Grandma ate very little. She had always been physically active but says now that she is immobile, she is simply not as hungry. Although this makes sense, her children worry that she is not eating meat or dairy products, and may become ill from this. She has been experiencing nausea, dizziness, and dry mouth at times, and urinating can be very painful. As a result she never takes more than two sips of her juice at breakfast or her water at lunch and dinner. The children decide she should see a doctor. She agrees and they take her to a local physician.

The doctor advises her to make an effort to return to her former good eating habits. He also found that she had a urinary tract infection, and was about to prescribe tetracycline when he discovered that she is taking Coumadin for heart disease and aspirin for her hip. He finds an alternative medication. The family hopes that with sound medical advice and time Grandma will return to her former healthy habits and be the same cheerful woman she always was.

Questions

1. Grandma is on many different medications. Describe some of the consequences drug interactions can have.

2. Why does the doctor quickly change his prescription of tetracycline?

3. What is the most likely cause of Grandma's urinary tract infection?

4. What possible interaction should Grandma be aware of regarding the Coumadin she is taking?

5. Although she claims she is simply not hungry, what could Grandma's lack of appetite be an indication of?

6. Grandma does not drink milk, but needs calcium in her diet. How much calcium does she need, and what can she do to add this to her diet without drinking milk?

7. Aside from calcium, milk is also a substantial source of vitamin D. Why is adequate milk consumption more of a concern than before, given Grandma's current health and recent change in living environment?

by Thomas W. Cas'

Case Study #18: Genetic Predisposition to Chronic Disease

Clyde, a 30-year-old marketing executive, has been reading a great deal lately about the Human Genome Project. Modern science is offering new opportunities to anyone who wishes to know about the genetic predispositions they may have to certain diseases. Clyde has learned that modern genetic screenings validate many people's predictions based on their family histories. Motivated by his recent reading, Clyde researches his family history and develops a list of possible health concerns.

Unfortunately for Clyde, the list of possible health complications is rather lengthy. All of Clyde's family members are overweight, including Clyde. Many of his relatives have high blood pressure. Hypertension led to his grandfather's heart attack, and two of Clyde's uncles have Type 2 diabetes. Clyde is most concerned about his genetic predisposition to colorectal cancer. Several cousins were diagnosed with this cancer while in their thirties. Clyde knows that thousands of people each year die from cancer, and decides that it is not too late to change his dietary habits in order to minimize his apparent genetic susceptibilities.

Clyde is now devoted to eating right and exercising. He restricts his caloric intake and exercises for 30 minutes 5 days per week. Clyde had always been slightly overweight, but now is very fit and able to maintain his ideal body weight. He also gets regular cholesterol screenings and limits the amount of meat he eats to help lower his cholesterol and saturated fat intake. Clyde used to drink wine or beer most nights at dinner, but now he no longer drinks more than three alcoholic drinks per week. He has never smoked, but now avoids even being around those who do. Clyde drastically limits the amount of salt he eats, while adding a multitude of fresh fruits and vegetables to his previously high protein, meat-based diet. He now eats fish regularly, and incorporates garlic or ginseng into most of his meals. Finally, he no longer spends his summer weekends working on his tan.

Questions

1. Which dietary practices may help Clyde prevent cardiovascular disease?

2. Which of Clyde's practices may help prevent hypertension?

3. How can diet and lifestyle help prevent cancer?

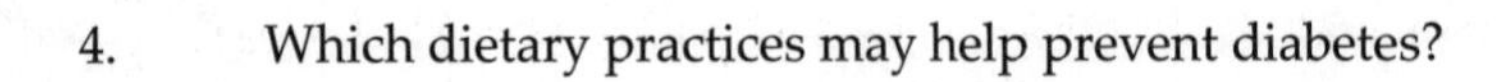

4. Which dietary practices may help prevent diabetes?

5. What herbal medicines does Clyde incorporate in his diet and what are their possible effects?

Case Study #19: Food Safety on the Go

Jessie lives in Wyoming, right outside of Yellowstone National Park, and spends much of her free time camping and hiking in the area. She is excited when her sister, Abbey, and Abbey's two children come to visit. Abbey's son Jordan is only a few months old, but her daughter Nicole just turned five. She is old enough to enjoy the outdoors, so Jessie invites Nicole to go hiking for the day while Abbey stays home with Jordan.

"What would you like to pack for lunch?" Jessie asked Nicole as they prepared for their hiking trip. "Cookies!" Nicole replied enthusiastically. Jessie packed cookies for dessert, tuna salad sandwiches prepared with mayonnaise, lettuce and tomato, and apples. She also packed plenty of bottled water. Jessie is always very hesitant to eat anything with pesticides or artificial coloring, so the lettuce and tomato on her sandwiches were grown in her own garden. The apples were organically grown, and they have the USDA sticker to prove it. Once she has placed them in an insulated backpack with an icepack, the two are on their way to experience the great outdoors.

They hike for hours, stopping by a stream to eat their lunches. "I'm thirsty," remarked Nicole as she headed to the stream to take a drink. Jessie stopped her and handed her some bottled water. After lunch they continued following the trail. Jessie kept a close eye on Nicole, who enjoyed picking berries she found on shrubs along the sides of the trail. Jessie thought they might be blackberries, but wasn't positive, so she did not allow the child to eat them. The two made it back home just in time to watch a beautiful sunset.

"Dinner is ready and waiting, " said Abbey as they walked in the door. Nicole and Jessie were glad to hear this because they were starving. Abbey had made baked chicken, steamed broccoli, and roasted potato wedges served with a honey dipping sauce made from raw honey and mustard. Everything looked delicious, but Jessie noticed that some of the potato wedges had a layer of green under the skin. Luckily Abbey had made more than enough potatoes, and Jessie knew enough to discard the wedges with green layers. Even Jordan seemed to enjoy the meal, although he was not allowed the honey dipping sauce. Nicole ate as much as she could and then excused herself to go to bed. She was exhausted after a full day of hiking with her aunt Jessie.

Questions

1. Jessie only eats organically grown fruits and vegetables. Is this necessary? What are some of the pros and cons of doing so?

2. Why was it so important that Jessie bring a cooler and ice pack?

3. Jessie promptly stops Nicole from drinking water out of the stream. Why would this water not have been safe? Luckily, Jessie brought bottled water, but if she had not what could she do to prepare the stream water so it would be safe to drink?

4. Was Jessie right to keep Nicole from eating the berries? Why or why not?

5. Abbey prepared potatoes that had green spots. What is this an indication of? Explain why Jessie felt it was important to make a new batch, even though the potatoes were cooked.

6. Jordan was not allowed to eat the honey dipping sauce. Why is this?

Case Study #20: Environmental Dilemma

Brette is an Environmental Studies major at her college in California. She is taking a class on the agriculture industry, and is very interested in understanding differences in farming techniques. As a class trip, the professor takes them on a tour of several farms in the area.

The first stop is in Apple Valley, right outside of Sacramento. The farmer, Glen, is an older man, who explains that he is from several generations of farmers. They have learned from one another and prefer to use traditional farming techniques. Glen has incorporated some newer technology, however. Water is often a scarce resource in California. Glen uses modern irrigation technology to conserve water in ways that were not available to his grandfather. He can focus water use during dry spells on specific areas of greatest need. His farm takes a lot of work, but he produces apples, all grown without the use of pesticides. Glen explains that the seeds he uses were genetically engineered. These "special" seeds have produced trees that are pest resistant and disease resistant, so pesticides aren't needed. The apples have a vibrant color and are better able to withstand storage and shipping. He is very proud of his ability to combine the traditional styles of his father and grandfather with environmentally safe practices available through modern technology.

Next is a visit to an almond orchard in Arbuckle, CA. The farmer, Maggie, prides herself on the sheer quantities she ships out every week. She explains that her orchard employs many workers that all get paid fair wages because the farm is so profitable. To protect the crops and ensure mass production, Maggie uses any means available to reduce pest populations. She employs mechanical trapping devices, pheromones, biological pesticides, and, if needed, chemical pesticides. Although this may involve some environmental risks, ensuring crop growth enables them to maximize the production of almonds.

The class now makes its way to Sacramento Valley. They visit a rice farm. Although large, this farm seems to have very few employees. The owner, a man named Bill, explains that he needs fewer employees than other rice farms this size, because modern technology contributes extensively to his farming practices. For example, he uses laser technology to precision level and grade fields, tractors to prepare seedbeds, and self-propelled combines for handling muddy soil. He uses global positioning satellites to pinpoint the specific parts of rice paddies that need attention. Although this technology is expensive, the farm saves money from lower labor costs. He can use less water, save fuel, and reduce the use of pesticides. Bill is able to adjust his use of resources to ensure the least waste and environmental harm, while keeping production at a maximum.

Finally, Brette and her classmates reach the Central Valley. They see acres and acres of cornfields. The farmer here is named Joseph. He explains that his corn can resist most anything. Not only is the corn designed to grow in drought conditions, but also it is "Round up" ready. "Round up" is a herbicide that he uses two or three times per year to kill the weeds that may threaten the crop. The corn itself is genetically resistant to this herbicide so although weeds are killed, the corn survives.

Once back at school the students must gather the information they have collected. They are required to classify each farm as either high-input or low-input, and compare the different farming practices used in terms of economic and environmental impact.

by Thomas W. Castonguay

Questions

1. In which classification would you place the first farm in Apple Valley? Why?

2. In which classification would you place the farm in Arbuckle? Why?

3. In which classification would you place the farm in Sacramento Valley? Why?

4. In which classification would you place the farm in Central Valley? Why?

5. Describe farming techniques common to several of these farms.

6. Describe the costs and benefits of each farming technique.

7. Among the farms discussed, which one do you think is the most environmentally friendly? Why?

by Julian H. Williford, Jr.

HOW TO RECORD YOUR 24-HOUR DIETARY HISTORY

1. Record **every item** that you consume in one 24-hour period.

2. Be sure to record:

   - The **amount consumed** (refer to your text's food composition tables to determine the amount of each specific food considered to be one serving, sometimes listed as "measure").
   - If you ate more than one serving, record as 1.5 or 2 or 3.75, etc.
   - If you ate less than one serving, record as 0.75 or 0.5, etc.

3. Record **how the food was prepared or cooked** (fried, broiled, poached, raw, etc.).

<u>Format</u>:

Name of Food	Amount Consumed	How Prepared
honey-nut cereal	2 cups = 2 servings	no preparation
fortified soy milk	1 cup = 1 serving	no preparation
whole-wheat bread	1 slice = 1 serving	toasted

4. Keep this list for 24 hours. Then enter this information into the *Diet Analysis Plus* computer program.

5. When you have entered all of your foods into the computer database, print your intake spreadsheet and complete set of other reports for that day.

6. Repeat above steps for the number of days you have been assigned. Print each of the following reports to see an analysis of your average intake over several days. To obtain an average, you should select the first day you recorded under "From date," and the last day you recorded intake as the "To date," for each of these reports:

   - Macronutrient Ranges
   - Fat Breakdown
   - Intake vs. Goals
   - Food Pyramid Analysis

THREE-DAY DIET ANALYSIS PROJECT

1. Format for report: typed, double-spaced.

2. Your report must contain the following **in order**:

   a. **Cover page** with title, name, date
   b. **Introduction:** One paragraph explaining why you are doing the analysis.
   c. **Results:** Printouts from Diet Analysis Plus: **(1)** the bar graph describing food consumption and the percentage of nutrient requirement being met for the average of 3 days, **(2)** the Food Guide Pyramid for the average of the 3 days and **(3)** the profile of the individual's requirements. (Do **NOT** include **individual** days' information.)
   d. **Discussion:** Answer the following questions in complete sentences.

      1. What was your **average caloric intake** for the 3 days? After 6 months of eating this way, how much weight will you gain or lose?
      2. How many **grams** and **calories** of carbohydrates did you consume? What **percentage** of your total calories came from carbohydrates?
      3. How much **total fat** did you consume in **grams** and **calories**? What **percentage** of your calories came from total fat?
      4. How much protein in **grams** and **calories** did you consume? What **percentage** of total calories came from protein?
      5. For those vitamins and minerals that you consumed in deficit of the RDA, **first list them** and **name one or several foods that you would HAVE to eat and the amount needed per day to meet the RDA for EACH of the vitamins or minerals in which you are deficient.** If you have **no** deficits, please address any of the following areas in which you are in **excess**: vitamin A, D, E, K, B_6, or iron.

Please **keep** all your information from Diet Analysis Plus, including each day's information as well as the averages. You may be asked to refer to it again.

by Jill S. Golden, Eleanor B. Huang, and Gail Meinhold

2005 *DIETARY GUIDELINES FOR AMERICANS* EVALUATION

Do not begin these worksheets until your 3-day *Diet Analysis Plus* printout has been approved by your instructor. Before submitting your printouts for approval, use the "Computer Project Self-Check" (p. 70) and make corrections if necessary. You should include the following reports: Profile DRI Goals, Macronutrient Ranges, Fat Breakdown, Intake vs. Goals, and Food Pyramid Analysis (print reports for each of the 3 days, and for the average of all three days).

This evaluation of your diet is based on the 2005 *Dietary Guidelines for Americans* and divided into 8 parts: *Adequate Nutrients within Calorie Needs, Weight Management, Physical Activity, Food Groups to Encourage, Fats, Carbohydrates, Sodium and Potassium, and Alcoholic Beverages.* In each section we will examine your diet to identify its strengths and weakness. After completion of the worksheets you will be asked to write a 2-page discussion.

Adequate Nutrients within Calorie Needs

- *Consume a variety of nutrient-dense foods and beverages within and among the basic food groups while choosing foods that limit the intake of saturated and* trans *fats, cholesterol, added sugars, salt, and alcohol.*

Using a highlight pen, highlight on your printout for the 3-day average the following.

- **Basic Components (Percent based on your personal profile):**
  - Calories below 80% or above 120%
  - Protein below 100% or above 200%
  - Carbohydrate below 130 grams
  - Fiber less that 80%
  - Fat and saturated fat above 100%
  - Mono and polyunsaturated fat below 80%
  - Cholesterol above 100%
  - Vitamins and minerals (except sodium and potassium) below 80% or above 200%
  - Potassium below 100%
  - Sodium above 100%
- **Source of Calories (Percent based on calories you consumed):**
  - Carbohydrate below 45% or above 65% of calories
  - Fat below 20% or above 35%
  - Protein below 10% or above 35%
- **Source of Fat (Percent based on calories you consumed):**
  - Saturated fat above 10%
  - Mono and polyunsaturated fat below 10%

You have just highlighted the areas of your diet that are above or below the recommendations established in the Dietary Reference Intakes (DRI).

Weight Management

- *To maintain body weight in a healthy range, balance calories from foods and beverages with calories expended.*

Estimate Your Energy Requirements

1. **Basal Metabolism:** Use the same weight that is on your computer printout. To determine the number of calories for your basal metabolism, convert your weight in pounds to kilograms.

   My weight = _________ lb. ÷ 2.2 = ____________ kg

   Basal Metabolism
   Basal requirement = 1.0 cal/kg/hr for men or
   Basal requirement = 0.9 cal/kg/hr. for women

   Body weight _______ kg × (1.0 men or 0.9 women) = _______ calories × 24 hours

   = ____________ Calories for Basal Metabolism (BMR)

2. **Activity calories:** Calculate all activity levels and record the same activity level you selected when you set up your *Diet Analysis Plus* profile below.

3. Sedentary Activity (20 %) BMR _______ × .20 = ________ Activity Calories

   Light Activity (30 %) BMR _______ × .30 = ________ Activity Calories

   Moderate Activity (40 %) BMR _______ × .40 = ________ Activity Calories

   Heavy Activity (50 %) BMR _______ × .50 = ________ Activity Calories

4. Thermic Effect of Food (TEF) is about 10% of the calories you consumed.

   3 Day Average for Calories ___________ × 0.10 = _____________ TEF

5. Your total estimated caloric expenditure is:

   __________ Basal Metabolic Rate

   \+ __________ Activity Calories, same as your Computer Project

   \+ __________ Thermic Effect of Food (TEF)

   = __________ Total Estimated Calorie Requirements (A)

   __________ Average calories you consumed over 3 days (B)

6. Compare A and B above. If there is more than a 300 calorie difference, which one (A or B) is a reasonable number of calories for you?

- *To prevent gradual weight gain over time, make small decreases in food and beverage calories and increase physical activity.*

Evaluate Your Weight

1. Determine your BMI using the chart:

Weigh yourself and have your height measured.

Weight ____________ Height _______________

Find your BMI category in the figure below. The higher your BMI category, the greater the risk for health problems.

BMI measures weight in relation to height. The BMI ranges shown below are for adults. They are not exact ranges of healthy and unhealthy weights. However, they show that health risk increases at higher levels of overweight and obesity. Even within the healthy BMI range, weight gains can carry health risks for adults.

Directions: Draw a line up from your weight at the bottom of the graph. Now draw a line across from your height, the point where you come to the line that matches your height is your BMI. Then look to see what weight group you are in.

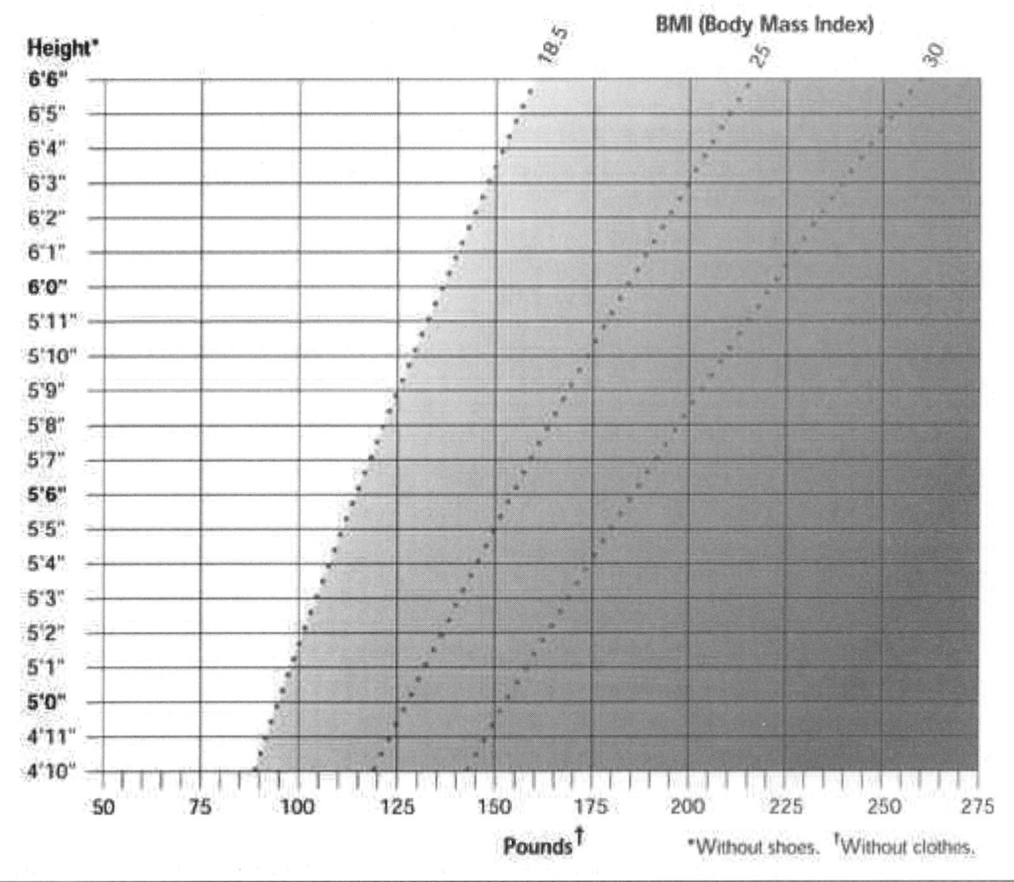

BMI Risk for Health Problems Related to Body Weight	
Underweight	BMI below 18.5 refers to underweight
Healthy Weight	BMI from 18.5 up to 25 refers to a healthy weight.
Overweight	BMI from 25 up to 30 refers to overweight.
Obese	BMI 30 or higher refers to obesity. Obese persons are also overweight.

2. My BMI is ________ and I am classified as ____________________________ from the chart above.

3. **Waist Measurement:** Measure around your waist, just above your hip bones, while standing. Health risks increase as waist measurement increases, particularly if waist is greater than 35 inches for women or 40 inches for men. Excess abdominal fat may place you at greater risk of health problems, even if your BMI is about right.

   Waist Measurement ______________

<u>Your Other Risk Factors for Chronic Disease</u>

- Do you have a personal or family history of heart disease? .... Yes/No
- Are you a male older than 45 years or a postmenopausal female? .... Yes/No
- Do you smoke cigarettes? .... Yes/No
- Do you have a sedentary lifestyle? .... Yes/No
- Has your doctor told you that you have
  - High blood pressure? .... Yes/No
  - Abnormal blood lipids (high LDL cholesterol, low HDL cholesterol, high triglycerides)? ... Yes/No
  - Diabetes? .... Yes/No

The higher your BMI and waist measurement, and the more risk factors you have, the more you are likely to benefit from weight loss.

Physical Activity

- *Engage in regular physical activity and reduce sedentary activities to promote health, psychological well-being, and a healthy body weight.*
- *Achieve physical fitness by including cardiovascular conditioning, stretching exercises for flexibility, and resistance exercises or calisthenics for muscle strength and endurance.*

Record your physical activity for the one week; include whether the activity was aerobic, anaerobic or both. Current recommendations suggest that you engage in at least 60 minutes of moderate physical activity most, preferably all, days of the week. Use only activities lasting more than 15 minutes:

Day	Activity	Number of Minutes	Aerobic	Anaerobic
Sunday				
Monday				
Tuesday				
Wednesday				
Thursday				
Friday				
Saturday				

Weight management and physical activity summary: Be prepared to discuss these in your report summary. Highlight the word *strength* or *weakness* below and answer why you chose this answer.

- Is your calorie intake a strength or weakness? Why?
- Is your BMI a strength or weakness? Why?
- Is your waist measurement a strength or weakness? Why?
- Do you have other risk factors? If so which?
- Is your activity level a strength or weakness? Why?

Food Groups to Encourage

Different foods contain different nutrients and other healthful substances. No single food can supply all the nutrients in the amounts you need. To make sure you get all the nutrients and other substances you need for health, adopt a balanced eating pattern, such as the USDA Food Guide or the Dietary Approaches to Stop Hypertension (DASH) Eating Plan.

Using each of the three days and your 3-day average Food Pyramid Analysis reports, list the number of servings you actually consumed.

Food Group	Day 1	Day 2	Day 3	3-Day Average	Within Range? Yes/No
Bread, Cereal, Rice & Pasta Group (6 – 11 servings; 3 whole grains)					
Vegetable Group (2 ½ cups from 5 sub-groups)					
Fruit (2 cups)					
Meat, Poultry, Fish, Dry Beans, Eggs & Nuts Group (2 – 3 servings)					
Milk, Yogurt, & Cheese Group (3 cups fat-free or low-fat)					
Fats, Oils, & Sweets (use sparingly)					

- *Consume a sufficient amount of fruits and vegetables while staying within energy needs.*

by Jill S. Golden, Eleanor B. Huang, and Gail Meinhold

Record 3-4 foods that you consumed during these three days in each category below. If you ate none, write none.

<u>**Vegetable Group**</u> (2 ½ cups)

Day 1: ______________________, ______________________, ______________________

Day 2: ______________________, ______________________, ______________________

Day 3: ______________________, ______________________, ______________________

<u>**Fruit**</u> (2 cups)

Day 1: ______________________, ______________________, ______________________

Day 2: ______________________, ______________________, ______________________

Day 3: ______________________, ______________________, ______________________

- *Consume 3 or more ounce-equivalents of whole-grain products per day, with the rest of the recommended grains coming from enriched or whole-grain products. In general, at least half the grains should come from whole grains.*

<u>**Bread, Cereal, Rice & Pasta Group**</u> (6 – 11 servings; 3 whole grains)

Day 1: ______________________, ______________________, ______________________

Day 2: ______________________, ______________________, ______________________

Day 3: ______________________, ______________________, ______________________

  - Use a highlight pen and highlight the whole grains in your list above. If none, what whole grains such as whole wheat, brown rice and oats could you add to your diet to improve your intake?

- *Choose a variety of fruits and vegetables each day.*

  - Use a highlight pen with a different color from the whole grains and highlight the fruits and vegetables minimally processed, eaten raw, or prepared using low fat methods.
  - If no foods were prepared minimally processed, eaten raw or prepared using low fat methods, what specific foods could you add to increase your intake?

  - Vitamin C intake from 3-day average = __________ mg, ________% goal

Look up vitamin C on your daily spreadsheets and record your 2 highest foods for each day.

Day 1: ______________________________ , ______________________________

Day 2: ______________________________ , ______________________________

Day 3: ______________________________ , ______________________________

If you were below 80% of your goal, what specific foods would you add to increase your vitamin C intake?

- Vitamin A (retinol & beta-carotene) intake from 3-day average = __________ RAE, __________ % goal

Look up Vitamin A on your daily spreadsheets and record your 2 highest foods for each day.

Day 1: ______________________________ , ______________________________

Day 2: ______________________________ , ______________________________

Day 3: ______________________________ , ______________________________

If you were below 80% of your goal, what specific foods would you add to increase your vitamin A intake?

- Vitamin E intake from 3-day average = __________ mg, __________ % goal

Look up vitamin E on your daily spreadsheets and record your 2 highest foods for each day.

Day 1: ______________________________ , ______________________________

Day 2: ______________________________ , ______________________________

Day 3: ______________________________ , ______________________________

If you were below 80% of your goal, what specific foods would you add to increase your vitamin E intake?

- Folate intake from 3-day average = __________ mcg, __________ % goal

Look up folate on your daily spreadsheets and record your 2 highest foods for each day.

Day 1: ______________________, ______________________

Day 2: ______________________, ______________________

Day 3: ______________________, ______________________

If you were below 80% of your goal, what specific foods would you add to increase your folate intake?

- Fiber intake from 3-day average = __________ g, __________ % goal

  Look up fiber on your daily spreadsheets and record your 2 highest foods for each day.

  Day 1: ______________________, ______________________

  Day 2: ______________________, ______________________

  Day 3: ______________________, ______________________

  Is the majority of fiber coming from soluble or insoluble fiber?

  Describe the benefits of both types of fiber (soluble and insoluble) and give 2 examples of each type of fiber that you are currently consuming or that you can add to your diet. If necessary, use a separate sheet for your answer.

  If you were below 80% of your goal, what additional foods would you add to increase your fiber intake?

- *Consume 3 cups per day of fat-free or low-fat milk or equivalent milk products.*

Milk, Yogurt, & Cheese Group (3 cups fat-free or low-fat)

Day 1: ______________________, ______________________, ______________________

Day 2: ______________________, ______________________, ______________________

Day 3: ______________________, ______________________, ______________________

- If you consumed less than 2 servings from the milk group, what foods would you add to increase your dairy intake?

- Would you want to do this? Why or why not?

Meat, Poultry, Fish, Dry Beans, Eggs & Nuts Group (2-3 servings)

Day 1: ______________________, ______________________, ______________________

Day 2: ______________________, ______________________, ______________________

Day 3: ______________________, ______________________, ______________________

- If you ate more than 4-6 ounces of meat (beef, lamb, pork) or poultry total each day, how could your meat, poultry, fish intake be decreased?

- How would this change affect your fat intake?

Fats, Oils, & Sweets (use sparingly)

Day 1: ______________________, ______________________, ______________________

Day 2: ______________________, ______________________, ______________________

Day 3: ______________________, ______________________, ______________________

- If you consumed more than 10 servings in this group, what specific foods could you reduce or remove to decrease your intake?

Calcium

Adolescents and adults over age 50 have an especially high need for calcium, but most people need to eat plenty of good sources of calcium for healthy bones throughout life.

Read the sections on osteoporosis and calcium in your textbook and write a summary of how each of the following factors may affect you. Write neatly (or type) in the space below and, if necessary, use a separate sheet for your answer.

by Jill S. Golden, Eleanor B. Huang, and Gail Meinhold

1. Your gender: What is the significance of gender as a predictor of osteoporosis?

2. Are you younger than 30? What is the important of age?

3. Do you regularly participate in weight-bearing exercise? Why is weight-bearing exercise so important?

4. Do you have a family history of osteoporosis? What is the significance of genetics and ethnicity as a risk factor?

5. Females, have you reached middle age or are you post menopausal? Why is this important?

6. Do you smoke and/or drink alcohol (more than 2 alcoholic drinks per day for males or 1 drink per day for females)? What are the associated bone risks for those who smoke or drink alcohol?

7. Do you take calcium supplements? Based on your calcium intake, do you need supplements?

8. How many grams of calcium do you consume daily?

   3-day average = _________ milligrams, _________ % of goal

9. If you consume less than 80% of your goal, what specific foods could you add to your diet to increase your calcium intake?

   Are you interested in making this change? Why or why not?

<u>**Iron**</u>

Young children, teenage girls, and women of childbearing age need good sources of iron, such as lean meats or cereals with added nutrients, to keep up their iron stores.

1. How many milligrams of iron do you consume daily?

   3-day average = _________ milligrams, _________ % of goal

2. Calculate the amount of this iron actually absorbed by multiplying the amount of iron consumed by .10 (10%).

   Intake = _________ milligrams × .10 = _________ milligrams absorbed

   - If you are a man over age 18 or a woman over age 50 you need to absorb 1 mg per day.
   - If you are a woman between 11 and 50 years old you need to absorb 1.5 mg per day.

3. Review the sections on trace minerals in your textbook and answer the following questions.

   - What effect will consumption of 3 ounces of meat/fish/or poultry (MFP Factor) have on your iron absorption?

   - What effect will being a vegetarian have on iron absorption?

   - What effect will consumption of 75 mg of vitamin C (ascorbic acid) or more at the meal including iron each day have on iron absorption?

- List 3 iron absorption-inhibiting factors:

  1.

  2.

  3.

- Does most of your iron come from iron-fortified cereals (hot or cold), energy bars or other fortified foods? Which of these do you consume and what percentage of iron is in each? (Refer to package labels for this information.)

  Food: ________________________________, ___________ % iron

  Food: ________________________________, ___________ % iron

4. If your iron intake is below 80% of your requirement, how might this affect you? (Refer to your book.)

   - What foods could you add to your diet?

5. If your iron intake is above 200% of your requirement, how might this affect you? (Refer to your book for information on iron toxicity.)

   - What foods could you remove from your diet?

Protein

3-day average protein intake = _________ grams, _________ % of goal

Write in this space any protein powders or amino acids supplements you consume each day in addition to what is reported in your diet analysis.

Protein powder: ______________________________, _________ grams protein per serving

Main protein ingredients as stated on the ingredient label:

Amino acid supplements: ______________________, ________ grams protein per serving

Main protein ingredients as stated on the ingredient label:

1. List 5 roles/function of protein in the body:
   - ____
   - ____
   - ____
   - ____
   - ____

2. What is the health effect of excess protein? (Can a high-protein diet harm you?)

3. Recommendations suggest that a safe level for protein is between 100% and 200% of your goal. If you consume less than 100% or more than 200% of your protein requirement, answer the questions below.

   - **Complete these questions if your protein is <u>below</u> 100% of your goal.** If your protein intake is generally below 90%, you possibly are not getting enough protein for your daily needs. Read the sections in your textbook on protein, especially the sections on proteins synthesis, roles of proteins, and protein metabolism; then answer the questions below.

   - What effect will this have on your nitrogen balance?

   - If your calorie intake is low, what function will the protein be used for?

   - What foods could you increase in your diet to increase your protein intake?

   - **Complete these questions if your protein is <u>above</u> 200% of your goal or you are taking any protein or amino acid supplements.** If you are eating this much protein you are spending protein prices for an energy-yielding nutrient and are displacing other important foods with too many protein-rich foods. If this is the case, read the sections in your textbook on protein metabolism, including the use of amino acids for energy; then answer the questions below.

- What is the fate of extra protein in the body?

- What implications might this have for your health?

- Why are you doing this?

- What specific foods could you decrease in your diet? (Use the spreadsheets for each day.)

Use of Dietary Supplements

Do you consume dietary supplements such as vitamins and minerals or herbal products?

1. Do you consume dietary supplements? This includes multivitamin and mineral supplements.

2. If you do not take supplements, does your diet necessitate them? Why or why not?

3. The supplements below are toxic when taken in excess (above the tolerable upper intake level). Review the supplements you take and complete the chart below for those nutrients.
   a. Record the dosage of the supplement as listed on the bottle.
   b. Record your intake (from foods/beverages) from the 3-day average.
   c. Record recommendation from your Profile DRI Goals report.

Supplement	Dosage	Your Intake from 3-Day Average	Your DRI Goal from Profile	Tolerable Upper Intake Level
Vitamin D	mcg*			50 mcg.
Vitamin E	mg			1000 mg
Iron	mg			45 mg
Zinc	mg			40 mg

* mcg and μg are both abbreviations for micrograms

4. If you consume dietary supplements, use the chart above and your printout, and review the information in your textbook on vitamin/mineral supplements, antioxidant nutrients, and phytochemicals, and their roles in disease prevention. Then answer the following question.

   - In light of the above what changes do you plan to make?

5. List any herbal supplements you take here:

6. Refer to the sections on herbal supplements in your textbook and summarize the pros and cons of the supplements you listed above. (Attach additional pages if necessary; type.)

Food groups to encourage summary: Highlight the word *strength* or *weakness* below to use in your summary. Include a short answer about why you think this is a strength or weakness.

- Is the variety of foods you consume a strength or weakness? Why?
- Do you consume whole grains? Strength or weakness?
- Do you consume a variety of fruits and vegetables? Strength or weakness?
- Is your calcium intake a strength or weakness? Why?
- Is your iron intake a strength or weakness? Why?
- Is your protein intake a strength or weakness? Why?
- Is your supplement use a strength or weakness? Why?

Fats

- *Consume less than 10% of calories from saturated fatty acids and less than 300 mg/day of cholesterol.*
- *Keep total fat intake between 20-35% of calories, with most fats coming from sources of polyunsaturated and monounsaturated fatty acids, such as fish, nuts, and vegetable oils.*

Fats supply energy and essential fatty acids, and they help absorb the fat-soluble vitamins A, D, E and K and carotenoids. You need some fat in the food you eat but choose sensibly.

What is your total fat intake? __________ grams, ________% goal

What is your saturated fat intake? __________ grams, ________% goal

What is your monounsaturated fat intake? __________ grams, ________% goal

What is your polyunsaturated fat intake? __________ grams, ________% goal

- *When selecting and preparing meat, poultry, dry beans, and milk or milk products, make choices that are lean, low fat, or fat free.*

List 3 of your highest food sources of fat for each day on the chart below. **Check (✓) the highest source of fat for each food.**

Food Source	Sat (✓)	Mono (✓)	Poly (✓)
Day 1			
1			
2			
3			
Day 2			
1			
2			
3			
Day 3			
1			
2			
3			

Using your 3-day average Fat Breakdown report, fill in the chart below. This chart represents the percentage of your total calories that come from each of the major fats and should add up to approximately 20-35% of your total calories.

Source of Fat	Recommendation	Your Intake	Within Recommendation? Yes/No
Saturated	10% or less		
Monounsaturated	10-15%		
Polyunsaturated	Up to 10%		
Other/Unspecified (includes *trans* fats)	---		---

If your saturated fat intake is above 10%, what foods might you reduce or replace to change this?

<u>Cholesterol</u>

How much cholesterol do you consume daily?

3-day average = __________ milligrams __________% goal

List cholesterol-containing foods you consume and the number of milligrams each contains.

Food	Cholesterol (mg)	Food	Cholesterol (mg)

- *Limit intake of fats and oils high in saturated and/or* trans *fatty acids, and choose products low in such fats and oils.*

<u>*Trans* Fatty Acids</u>

Trans fatty acids are not included in your printouts. *Trans* fats mimic the negative health effects of saturated fats. How many foods do you consume that contain *trans* fatty acids? Highlight those foods on the list below.

Margarine (hard, stick, soft, tub)
Shortening
Peanut butter
Pizza
Fried fast foods
Salad dressings
Mayonnaise
Biscuits
Rolls
Cakes
Pastries
Imitation cheeses
Crackers
Corn snacks
Corn chips
Potato chips
Other fried snacks
Cookies
Doughnuts
French fries
Deep-fried chicken
Deep-fried fish
Pies
Coffee creamers

1. If you highlighted 3 or more foods on the list above, what foods might you replace, decrease or eliminate to improve your diet?

2. Read the sections of your textbooks on the health effects and recommended intakes of lipids and on heart disease and strokes. What are the major risk factors for CHD (Coronary Heart Disease)? Highlight any you may have.

   o

   o

   o

   o

   o

   o

Fats summary: Reviewing this section on fat intake, answer the following questions:

- Is your total fat intake above 100% of your goal?

- Is your total fat intake below 50% of your goal?

- Is your saturated fat intake above 100% of your goal?
- Using your Fat Breakdown report, is your intake outside recommendations?
- Is your cholesterol above 100% of your goal?
- Do you consume more than 3 foods high in *trans* fatty acids per day?
- Is there a family history of cardiovascular disease in your family?
- Who?
- What effect does your intake have on your risk of cancer?
- What effect does your intake have on your risk of obesity?
- What effect does a very low intake of fat have on your health?
- What changes might you make in your diet as a result of what you have learned?

Carbohydrates

- *Choose fiber-rich fruits, vegetables, and whole grains often.*
- *Choose and prepare foods and beverages with little added sugars or caloric sweeteners, such as amounts suggested by the USDA Food Guide and the DASH Eating Plan.*
- *Reduce the incidence of dental caries by practicing good oral hygiene and consuming sugar- and starch-containing foods and beverages less frequently.*

The *Dietary Guidelines* urge people to choose beverages and foods that limit added sugar intake. Specifically, sugars should account for only 10 percent or less of the day's total energy intake. Food labels list the total grams of sugar a food provides. This total reflects both added sugars and those occurring naturally in foods. You can tell which foods are high in added sugar by checking the ingredient list on the food; if it starts with several types of sugar than the food is high in added sugar.

Recommendations suggest that "added sugars" as listed above should comprise less that 10 percent of the day's total calories, based on studies that show a lower intake of essential nutrients in diets high in added sugar. Compute the number of calories this would represent in your diet.

Total kcalories from 3-day intake average = __________ × .10 = __________ 10% of total kcalories

Review each day and **highlight** the foods consumed that are high in **added sugar** on the chart below.

Products high in added sugar	Approximate grams of sugar	# of teaspoons	kcal from sugar
Sweetened fruit drinks, carbonated and non-carbonated (12 ounces)	48	9	192
Cola (12 ounces)	36	7	144
Sherbet (½ cup)	28	5	112
Lemonade (8 ounces)	24	5	96
Fruit punch (8 ounces)	24	5	96
Candy (assorted, 1 ounce)	20	4	80
Cookies, commercial (4-5 small)	20	4	80
Donuts, yeast, glazed (1)	20	4	80
Gelatin desserts (1 cup)	20	4	80
Pastries, pan dulce (1)	20	4	80
Pie (1/6 of 9″ pie)	20	4	80
Popcorn, caramelized (1 cup)	20	4	80
Tang (6 ounces)	20	4	80
Ice Cream (1 cup)	20	4	80
Popsicle (1)	16	3	64
Granola bar (1)	16	3	64
Fruit rolls (1 fruit leather roll)	12	2	48
Brown sugar (1 tsp)	5	1	20
Candy (1 tsp)	5	1	20
Honey (1 tsp)	5	1	20
Jam or jelly (1 tsp)	5	1	20
Maple sugar or maple syrup	5	1	20
Molasses (1 tsp.)	5	1	20
Catsup (1 Tbsp.)	5	1	20

Review the sections on health effects and recommended intakes of sugars in your textbook and answer the following questions:

1. Does your intake of sugar create nutrient deficiencies in your diet?

2. Does your intake of sugar contribute to other problems such as obesity, dental caries or heart disease?

3. What foods would you replace or remove because of this exercise?

Sodium and Potassium

- *Consume less than 2,300 mg (approximately 1 teaspoon of salt) of sodium per day.*
- *Choose and prepare foods with little salt. At the same time, consume potassium-rich foods, such as fruits and vegetables.*

In the body, sodium – which you get mainly from salt – plays an essential role in regulating fluids and blood pressure. Many studies in diverse populations have shown that a high sodium intake is associated with higher blood pressure.

Salt (sodium chloride) is the main source of sodium in foods. Only small amounts of salt occur naturally in foods. Most of the salt you eat comes from foods that have salt added during food processing or during preparation in a restaurant or at home.

Your intake of sodium is __________ mg, __________ % of goal

Complete the table with the 4 foods that contributed the most sodium to your diet for each day.

Food Source	Milligrams
Day 1	
1	
2	
3	
4	
Day 2	
1	
2	
3	
4	
Day 3	
1	
2	
3	
4	

To get an idea of how much sodium is found in common foods we eat complete the chart below using a food composition table, labels, or *Diet Analysis Plus* software.

Food	mg of sodium
1 tsp salt	
10 potato chips (1 ounce)	
Chicken noodle soup, dehydrated, prepared with water, 1 cup	
Catsup, ¼ cup	
Soy sauce, 1 tsp	
McDonald's Quarter Pounder with cheese	
Pizza Hut, Personal Pan Pizza Supreme	
Taco Bell, taco salad with salsa	
Taco Bell Big Beef Burrito Supreme	
Oscar Mayer Lunchable, ham and cheese	

1. If your sodium intake is high (greater than 100% of goal), read about sodium in your textbook and discuss the possible implications for your health.

2. In what ways can you reduce your sodium intake? List specific foods you could reduce, replace or eliminate.

Your intake of potassium is __________ mg, __________ % of goal

Look up potassium on your daily spreadsheets and record your 2 highest food sources for each day.

Day 1: ____________________________, ______________________________

Day 2: ____________________________, ______________________________

Day 3: ____________________________, ______________________________

3. If your potassium intake is less than 100% of goal, in what ways could you increase intake? List specific foods you could replace or add.

Alcoholic Beverages

- *Those who choose to drink alcoholic beverages should do so sensibly and in moderation.*

Moderation is defined as no more than one drink per day for women and no more than two drinks per day for men. This limit is based on differences between the sexes in both weight and metabolism.

One drink is equal to:
- 12 ounces of regular beer (150 calories)
- 5 ounces of wine (100 calories)
- 1.5 ounces of 80 proof distilled spirits (100 calories)

Do you usually exceed this recommendation? If so, comment on the effect alcohol may have on your health.

Summary for fats, carbohydrates, sodium and potassium, and alcoholic beverages: Review these sections and answer the following questions. Be sure to include an explanation for each question.

- Is your fat, saturated fat, *trans* fat, and cholesterol intake a strength or weakness? Why?

- Is your sugar intake a strength or weakness? Why?

- Is your sodium intake a strength or weakness? Why?

- Is your alcoholic intake a strength or weakness? Why?

Written summary of your diet strengths and weaknesses

Now that you have reviewed the important elements of your diet, write a two-page evaluation of the most important strengths and weaknesses you identified in each section above. Type double-spaced.

1. Review the summaries at the end of each section noting the major strength and weaknesses.

2. Compose a separate paragraph for each section including the following:
    - Discuss the major weakness and the effects it may have on your health.
    - Address each weakness with recommendations for foods that can be eliminated, replaced, or added.
    - Discuss the major strengths with examples of successful strategies you use.

3. Close with a plan you will implement for the improvement of your diet. Make sure your plans list at least 5 **specific** goals you will implement to improve your diet and health.

4. For the last paragraph please help us improve the project by sharing with us your feeling(s) on this assignment and suggestions for improvements.

Thank you! We hope you found this exercise useful and we hope your nutritional status benefits from this evaluation of your current food choices.

COMPUTER PROJECT SELF-CHECK

Before submitting your 3-day diet record and computer printout, check your data for accuracy by searching for the items below. After making all corrections, get a new 3-day average print-out. Turn in: (1) this sheet, with (2) your 3-day food record, and (3) your corrected computer printout, stapled in the order shown in the instructions, with (4) the grading sheet (p. 71), all in a (5) 2-pocket folder.

Item	Yes	No	Corrected?	Comments
Did you record all beverages throughout each day and evening?				
Check the liquids – did you measure them accurately in ounces?				
Did you add sugar/creamers etc. to your hot beverages if you had them?				
Is your orange juice, lemonade, iced tea "made from concentrate" vs. concentrate? See calories – they should not be unusually high.				
Did you include margarine or other spreads on your bread, rolls, muffins?				
Were your chips or other snack items measured correctly? Bags at Subway, Quiznos, etc. are 1-1½ ounces.				
Is your rice or pasta cooked? (vs. dry!) This makes a huge difference in nutrients shown. Correct your data.				
Did you include salt, soy sauce, fish sauce, or other condiments (mayonnaise, ketchup, mustard, salad dressings, dips, salsa) each day in the correct amounts?				
Check your salads – are they measured in cups rather than ounces?				
Now look at your Profile – Compare those calories with the 3-day average calories you ate – are they close in number? (If calories in what you ate are much higher, see where they came from by looking at day 1, 2, 3 calories, e.g.1 loaf of bread vs. 1 slice!) (If calories you ate are much lower, you may not have included all your snack items, or put in the correct amounts.) You may need to correct your data so calories are closer to the Profile calories.				
Look at your Profile again – Compare your protein with the amount you ate in your 3-day average. Is the amount more than 200%? You may have over-estimated the amounts you ate – check and correct, as needed. **Do not** include protein supplements/ shakes/ vitamins/herbals in this food record.				

DIET ANALYSIS GRADING SHEET

Assignment	Possible Points	Your Points	Comments
Computer Printouts Must be stapled together correctly, in folder. Include this sheet and 3-day handwritten intake forms.	10		
Adequate Nutrients within Calorie Needs	5		
Weight Management Estimate your energy requirements	3		
Evaluate your weight	3		
Physical Activity Weight management/physical activity summary	3 3		
Food Groups to Encourage Food groups	7		
Calcium and iron	5		
Protein	5		
Use of dietary supplements	5		
Summary	3		
Category Recommendations Fat	5		
Sugar	5		
Salt, potassium and alcohol	5		
Summary	3		
Written Report	25		
Neat and in folder correctly	5		
Total:	100		

Place in a two-pocket folder the following:

- This grading sheet
- Food record input forms (handwritten record of what you ate)
- *Diet Analysis Plus* printouts
- Worksheets
- Written report

1-Week Diet Analysis: Actual vs. Model Diet

Format Issues:
Your diet analysis must be typed on unscented plain white paper, double-spaced, and stapled.

All written material submitted for credit must be typed and stapled. Use left margin justification, making sure your margins are no more than 1 inch on all sides. Use 12 point Times New Roman type or the equivalent. Use black ink. Do not use all CAPITAL letters for any portion of this assignment. No boldface or italic type. It is strongly suggested that you make a copy of all of your materials for yourself.

Part I (20 points):
Maintain a food diary for a single week. Write down the amounts of every food and beverage that you have consumed including snacks and fluids. Try to jot it down as soon as you've finished your meal or snack so you don't forget to record it. Record both the day and the meal (e.g. breakfast, lunch, dinner, snack) at which an item was consumed.

Record your impressions of your diet. Do you think you consume a healthful diet? What do you perceive to be the strengths and weaknesses of your diet?

Part II (40 points):
Use *Diet Analysis Plus* to analyze the diet you recorded for Part I. The printout of this diet analysis should be included in your report.

- Based on this diet, write a report that addresses the following points; please keep your response brief. Your response to these points may require doing analysis beyond that provided by the data analysis program.

  1. How does your weekly average energy consumption compare with current recommendations? This information may be presented in tabular form.
  2. What is your BMI? Does your BMI place you in the underweight, of appropriate weight, overweight, or obese group? Considering the weaknesses of the BMI, do you feel that your grouping is appropriate?
  3. How does your weekly average consumption of carbohydrates, proteins, and lipids compare (on both a percentage of kilocalories and absolute amount basis) with current recommendations? This information may be presented in tabular form.
  4. How does your weekly average consumption of simple sugars, complex carbohydrates, and fiber compare to current recommendations? If the diet analysis program does not allow you to distinguish between simple and complex carbohydrates you can group them together in your analysis. This information may be presented in tabular form.
  5. As both a percentage of kilocalorie intake and in terms of absolute amounts consumed, how does your weekly average consumption of cholesterol, polyunsaturated fats (separate omega-3 and omega-6 fatty acids from any other polyunsaturated fatty acids), monounsaturated fats, saturated fats, and *trans* fats compare to current recommendations? This information may be presented in tabular form.
  6. Discuss your intake of alcohol and discuss the benefits and risks of such a level of consumption.

Part III (40 points):

A. What specific recommendations can you make for improving your diet (i.e. specific foods that might be added to or deleted from your diet)? If you suggest deleting foods, what foods, if any, should take their place in your diet? If you suggest adding foods, what foods, if any, should be deleted from your diet?

B. Create a hypothetical diet based on your suggestions for improvement. Keep in mind that this hypothetical diet must consider that you do not want to consume all of the same foods every day. Analyze this diet based on the points listed in Part II as well as the additional points listed below. Repeat until you have created a healthy diet. Include the healthy diet and its analysis in your report.

7. For each vitamin, how does your weekly average consumption compare to recommendations? This information may be presented in tabular form.

8. For each mineral, how does your weekly average consumption compare to recommendations? This information may be presented in tabular form.

9. How does your consumption of water compare to current recommendations?

by Ruth L. Kuzmanic

HEALTHY HEART ANALYSIS

Are you at risk for a chronic disease? The work you do for this project will help you answer that question. One's wellness depends on many factors, some controllable and others not. You will assess your risk of chronic disease by analyzing these risk factors. Controllable risk factors include one's lifestyle choices of diet, exercise and stress management. Family health history is the predominant risk factor you cannot control. All of these risk factors will be addressed in your Healthy Heart Analysis.

The Healthy Heart Analysis requires Internet access. Anytime you use the Internet you may run into snags. Do your best to work through the snags, then ask your classmates and instructor for assistance.

Diet Analysis:

You will use the site www.FitDay.com to get the data for your diet analysis. When you visit the site for the first time, sign up and create an account. Then you can begin your diet analysis.

Directions to navigate FitDay.com:

I. After you are logged in, select **Set Goals**. From the menu on the left, create your nutritional goals by selecting the nutrient and clicking on **Go!** Set the minimum and maximum intake. Use the guidelines below or from your instructor to set goals for the following nutrients;

Total Fat:	30% of your daily calories
Saturated Fat:	10% of your daily calories
Total Carbohydrate:	60% of your daily calories
Protein:	10% of your daily calories

Dietary Fiber:	20-35 grams
Calcium:	1300 mg
Iron:	females = 15 mg; males = 12 mg
Sodium:	2400 mg

II. From the menu on the left, select **Guidelines**. Look under **Eat a Variety of Foods**, and select **"What counts as a serving?"** Read this prior to doing your diet analysis.

III. Now analyze an typical day's diet for yourself. Do this by clicking on **Foods**.

1. Enter a food you ate to search for.
2. Choose a food that closely resembles what you ate. You may have to substitute foods here to coincide with FitDay's data base. Click on **Add**, complete servings and serving size. Save changes. Print out the list of the foods that you ate by selecting **File print**.

IV. Print the following reports by clicking on **Reports**. Choose the following reports;

1. Am I meeting my nutrient requirements? (choose past week)
2. Am I meeting my custom nutrition goals? (choose past week)

V. Analyze an additional day's diet by changing the diet date and following the process again.

Energy Expenditure:

I. Determine the amount of calories you have burned. Again use the site www.FitDay.com.

Directions:

1. From the menu on the left, click on **Activities**.

2. Browse the activities, select one you have done on the same day as the food you are analyzing. Click **Go**.
3. Select the appropriate activity and click on **Add**. Complete the **Time Spent**.
4. From the menu on the left, select **Reports**. Select **Where did I burn my calories? (past week)** and print it.

Fast Food Analysis:

Visit the site www.foodfacts.info/ to analyze the basic nutrients found in meals at two different fast food restaurants. Your analysis should include data on calories, fat, carbohydrate, protein and sodium. Be sure to total each of the nutrients for the complete meal.

Directions:

1. Search by Restaurant. Complete the search fields.
2. Find the facts.
3. Include all the items you would eat for a meal.
4. Copy and paste into Microsoft Excel or Word to turn in with your completed project.

Family Health History:

Use the site http://generationalhealth.com/ to examine your genetic risk of chronic diseases. Be sure to include any incidences of heart disease, stroke, cancer, or diabetes that have occurred in your family. Begin by clicking **Create a New Family Tree** and follow the instructions on screen.

1. Include the following family members in your analysis: grandparents, parents, your parent's brothers and sisters, your brothers and sisters, and yourself.
2. In the event that there is no incidence of disease (remember, high blood pressure is a disease), simply list the required relatives and their ages. In the case of a deceased relative, list the age and cause of death as it relates to our study of chronic diseases.

Data Analysis:

Now that you have gathered all this interesting data it is time to analyze it to draw conclusions and make predictions regarding your wellness.

Address each of the following in your typewritten analysis:

1. A chart which compares your intake to the DRI for each of the 8 nutrients you have set a goal for.
2. A chart that includes data for each of the two fast food meals. Include calories, total fat, protein, carbohydrate and sodium for each item and totals for each of the meals.
3. Based on your diet analysis, write three measurable goals that would enhance your physical wellness. Be specific, with accountability and a time reference.
4. What diet changes do you need to make to meet or maintain your DRI for the each of the nutrients in section I?
5. Discuss your current energy expenditure and activity level in relation to your wellness. Are you going to gain weight, lose weight or maintain?
6. Examine your printout of your family health history. What diseases are you at risk for?
7. Now that you have analyzed your diet, energy expenditure, and family history, describe the lifestyle choices to avoid the chronic diseases.

by Chengshun Fang

SNACK ASSIGNMENT: THE *VIRTUAL SNACK SHOP*

DESCRIPTION:
Suppose a female student is looking for a low fat snack. She wants to have that food item to provide at least one vitamin and one mineral in significant amounts. Can you help her?

First, visit the Internet site for Dr. Richard Fang's *Virtual Snack Shop* at http://napa.ntdt.udel.edu/ntdt401/food/menu.asp. Browse the database by selecting several different vitamins and minerals to see some examples of wholesome snacks containing those nutrients like the snack you will be planning for this assignment. Your objective is to propose a new snack, different from all the snacks already available in the *Virtual Snack Shop*, which is rich in at least one vitamin and mineral.

CONDITIONS AND PROCEDURES:
1. Total energy for a day is set at 2000 kcal.
2. For the purpose of determining DRI, use female as the sex and 25 as the age.
3. Choose one of the following vitamins: vitamin C, thiamin, riboflavin, niacin, B_6, folic acid, B_{12}, A, D, E or K.
4. Choose one of the following minerals: calcium, magnesium, iodine, iron, and zinc.
5. Write down the name of the vitamin and the mineral and identify the DRI of each.
6. Identify a snack item based on criteria listed below.

   Criteria:
   1. This item should not be repeated from any of the earlier snacks listed in the *Virtual Snack Shop*, unless you choose a different set of vitamin and mineral. An item similar to the existing snack is also subject for a penalty. For example, if a cup of 2% milk was submitted earlier for calcium and vitamin D, a snack of one cup of 1% milk for calcium and riboflavin would be considered as not original. Browsing and learning from the earlier submissions through the *Virtual Snack Shop* is an important part of this assignment.
   2. The snack has less than 30% of energy from its fat.
   3. The snack has to provide no less than 30% of a vitamin's and 30% of a mineral's recommended dietary allowance (DRI) for a 25-year-old female with a portion size to provide 200 kcal. To verify this criteria, please follow the detailed instruction given in step #8.
   4. A snack normally should be no more than 400 kcal in energy, although the size of a portion will not change its nutrient density.
   5. A snack can be a combination of food items. In that case, please specify description and amount for each ingredient used in the snack.

7. The nutrient analysis of the snack can be accomplished by using diet analysis software or a food composition table. Be sure to turn in a copy of a proof of this analysis. The proof could be a computer printout or copy of a part of a food composition table. To ensure the accuracy of the analysis, make sure the following pieces of information are included and are highlighted with a color marker.
   - your name and date of completion
   - the name of the snack (or names of items included in the snack)
   - the portion size
   - total energy in kcal
   - % of energy from fat
   - the amount of the vitamin
   - the amount of the mineral

8. To determine if snack item is a significant source of the nutrient, Calculate the nutrient content for the snack as equivalent to a portion providing 200 kcal (10% of energy need) as follows:

   A. Divide 200 by the total energy of the snack (200/E);
   B. Multiply the above result by the Vitamin content of snack (V);
   C. Divide the result by the DRI value of the vitamin to get Index V.

   In summary, Index V = [V(200/E)]/DRIv

   If Index V is equal to or greater than 0.3, this snack is considered as a significant source of this vitamin because if this woman spends 10% of her energy allowance on this snack, she will receive 30% of the vitamin recommendation. (Note: Please keep two digits after the decimal points to ensure the accuracy.)

9. Repeat the above step for the mineral to figure out the Index M, which also has to be equal to or greater than 0.3.
10. If one or both nutrients failed to meet the criteria, redo steps 3 to 9 with adjusted recipe, or a set of micronutrients, or a different snack until a satisfactory item is found.
11. Then use the snack analysis form (following page) to turn in your report. Your recommendation will be reviewed by your peers as well as be recorded for grading. Double check your calculation to ensure the accuracy before submitting, and attach proof of the nutrient analysis (as described above) to the completed form.

SUBMITTING YOUR SNACK TO THE *VIRTUAL SNACK SHOP*:

12. When your instructor grades your assignment, he/she will let you know whether you have approval to submit the snack to the *Virtual Snack Shop*.
13. If your snack is approved for submission, use your completed assignment to fill in the on-line submission form for the *Virtual Snack Shop* at http://napa.ntdt.udel.edu/ntdt401/food/submit.htm.

Congratulations! You have helped yourself and other students learn how to choose nutrient-dense snacks for a healthy lifestyle.

TO LEARN MORE ABOUT THE *VIRTUAL SNACK SHOP*:

C. Fang, Virtual Snack Shop, An Internet Project with Emphasis on Food Sources of Vitamins and Minerals. *J Nutrition Education*. 2000, 32:119C.

SNACK ANALYSIS FORM

Be sure that you have read the instructions and completed the calculations before submitting your results. When you report a value, you need to include its unit. Attach proof of analysis to this form.

Name: ____

Vitamin chosen: ___________________________ DRI for chosen vitamin: _____________________

Mineral chosen: ___________________________ DRI for chosen mineral: _____________________

Food name and description:

____

____

____

____

____

____

Portion size and unit (such as in oz. or cups) = _______________________________

Type of nutrient analysis:
- ☐ computer diet analysis software
- ☐ food composition table
- ☐ food label

Energy = _________________ kcal % of energy from fat = _________________

Amount of the vitamin = _________________ Amount of the mineral = _________________

Index V = _________________ (It has to be 0.3 or greater.)

Index M = _________________ (It has to be 0.3 or greater.)

Before turning in this form, double check to ensure you have filled in every blank and included units when requested.

Feed Back: Do you think this is a good exercise and why?

DIET PROJECT: COMPARISON OF TWO DISTINCT DIETS

This project is to be completed using *Diet Analysis Plus* software. Data can be entered by food names.

Useful References:

Byrd-Bredbenner, *Nutr. Today*, 23(5):13, 1988.
Basiotis, Welsh, Cronin, et al., *J. Nutr.*, 117:1638-1641, 1987.
Karleck, *J. Am. Diet. Assoc.*, 87:869-871, 1987

OVERVIEW:

Choose a focus, an objective, for your project - some nutrition-related question to explore. This should involve you analyzing two diets which might be expected to be different and interpreting the results - noting good points, recommending improvements, commenting on surprises, stressing where expectations have been met, etc. Also discuss strengths and weaknesses of, or critically assess, the project. Your diet project must be scholarly.

Students in the past have explored typical diets at home vs. school, what's eaten while working in a foreign country vs. North America, what was consumed during the week in a shared student apartment of males vs. that in the house of his fiancé's mother on the weekend, a turn of the 20th century diet (grandma's diary) vs. a modern one or a comparison of popular weight loss diets.

PROCEDURES FOR THE DIET PROJECT

Before You Start:

Write out what the goal or focus of your study will be. Plan to study only diets which people are already consuming: you are not to actually "put" real people on any sort of diet. Consider who might record a 3-day diet history for you. These are normally close acquaintances. Obtain approval for your project from the instructor. This is best done by e-mailing the description of project and include a brief note on whom you intend to approach to participate: school mates, brother or sister, parent, team member, family friend, etc. All participants must sign the unaltered informed consent form (p. 86). The signed, dated, and witnessed consent forms must be included in your report or the project will not be marked. If the participant is a minor, a parent or guardian must sign the informed consent form and it must be witnessed by the course instructor in his office. Approval must be obtained before you can start your data collection. You will receive approval to start your project by return e-mail. Office appointments can be made of course.

The project's marking scheme is available to help prepare your report.

THE PROJECT:

Step 1:

Prepare 3-Day Diet Histories (be representative, e.g., 2 weekdays, 1 weekend day):

- Ask your participants to volunteer. Show them your written goals and explain to them what is expected and involved.
- Obtain their informed consent by their signature on the form (p. 86) and witness it.
- Get a description of them: height, weight, age, activity level, if dieting, pregnant or nursing, etc.
- NOTE: Code participant identities in all reports.

- Have your participants record all they eat at the time they consume it: method of preparation, amounts/size, kind, added foods, recipes, type of syrup/dressing, condiments, supplements, etc.
- Basic foods are easier to record/find/code/enter: watch out for altering the diet to accommodate the specific items available in the database.
- New foods can be entered using the information from the nutrition label.
- A recipe made up of foods already in the database can also be created, and the software will generate its nutrient contents based on the foods it contains.
- A personal activity history can be kept to estimate energy expenditure.

<u>Step 2</u>:

You may search for food items by name using *Diet Analysis Plus*. (Take a quick look at the *Diet Analysis Plus* program at the start of your project to become familiar with it.)

Selecting food items:

- In the food database are many combination foods, sandwiches and fast-food menu items which simplifies things.
- Some food substitutions will likely be necessary: explain your reasoning.
- Activity levels may be general or personalized, following the on-screen instructions in the Select and Edit Profiles window.

<u>Step 3</u>:

- Enter your data:
- Allow about 20 min, with a diet record of 20 - 25 items per day.
- Save your data just in case.
- Clicking on Create Reports and then on each of the reports available will show the nutrient content and other information about diet inputted for that day.
- Select the desired report – Profile DRI Goals, Macronutrient Ranges, Fat Breakdown, Intake vs. Goals, Source Analysis, Food Pyramid Analysis, or Intake Spreadsheet.

<u>Step 4</u>:

Many useful reports are available to help in the analysis stage of the project.

Printing:

- Reports are printed by displaying the desired report, then clicking the Print button in the upper right corner of the report window.
- To print a report for the average of the 3 days, select the desired report, and select the start and end dates under "From date" and "To date." Then, click the OK button.
- **Be sure to print the following reports** for each participant, as the minimum to attach to your report as an appendix: Profile DRI Goals; 3-day averages for Macronutrient Ranges, Fat Breakdown, Intake vs. Goals, and Intake Spreadsheet.

<u>Step 5</u>:

Analyze the diets and prepare your report of 2,500 - 3,000 words, excluding the appendix:

- State your goal or purpose, i.e., looking at two contrasting diets, the results obtained by different analysis methods, etc.
- Point out what's good and poor, make recommendations, suggest improvements, what's expected or surprises, which foods supply which nutrients, etc.

- Be critical: how good are estimates of serving size, the database, missing foods and data, ease of substitution, etc.
- Remember that the foods in the database never existed: they are averages of many types of varieties and of many individual samples of the food.
- Use the project to extend your knowledge in your field and interests.
- Be current, academic, topical.
- Focus your topic, show understanding of the data, new aspects and insights.
- Do not be too general. Use a general statement and a specific example to support your statement.
- Use in-text reference support for statements and a short bibliography.
- Use any generally accepted scholarly format.
- Your instructor will be looking for critical analysis, synthesis, evaluation skills, and insight.

Hand in your computer output – the summary tables (personal description and 3-day average including the energy and fat sources and ratios in particular) in particular - as an appendix.

- Clearly identify your work: use a distinctive cover for easier retrieval.
- The diets which you analyze and report must be unique to your project and the course.
- The majority of your references must be current and published within the last 5 years or so.
- Neatness, clarity, and organization, including the appendices' tables, are appreciated!

AFTERWARDS:
Give your participants a copy of the computer output of their own diet, share with them some of your insights and thank them for their help.

DIET PROJECT MARKING SCHEME

- informed consent forms included
- cover page clearly identifies project

Introduction and Methods

- goal/purpose clearly outlined; why did they choose this topic?
- method of analysis briefly outlined (recorded, entered and analyzed with *Diet Analysis Plus*)
- participants described:
  - sex; height; weight; activity level
  - other special conditions
- participants differentiated; how are the participants different?
- method of collecting data outlined: the 3-day dietary record method

Results

Each Individual Participant's Diet

- appendix includes tables on:
  - participant profiles
  - 3-day averages
- states nutrients significantly less than the DRI and/or greater than the DRI and outlines criteria for deciding what is significantly less than/greater than the DRI, e.g. <75% DRI
- good/bad aspects
- suggested improvements: be specific with examples – "reduce fat by replacing whole milk with skim"
- notes where expectations have been met
- outline any surprise results

Comparison of Participants' Diets

- similarities
- differences
- factors which may explain differences/similarities

Discussion

- must provide correct interpretation of data
- demonstrates an understanding of the DRI, and its limitations when applied to an individual's diet
  - DRI/RDA
    - the level of dietary intake thought to be sufficiently high to meet the requirements of almost all individuals in a group with specified characteristics
    - set for specific group of individuals, age and gender specific
    - based on population parameters
    - average requirement + 2 standard deviations (except energy)
    - thus, exceeds the requirement of almost all individuals
    - refers to average need over time, recommended amount does not have to be consumed each day, but on average
    - levels of intake are intended to maintain health in already healthy individuals
    - does not allow for illness/stress
    - based on typical dietary pattern of the country (culturally specific)
    - thus, may not be appropriate for atypical diets

        - assumes requirements for other nutrients are met
        - cannot be used to diagnose disease and/or deficiency - need clinical data
        - only provides an estimate of risk of nutrient deficiency; individual requirements are not known unless determined biochemically
- shows an understanding of 3-day dietary record method
    - strengths:
        - does not rely on memory
        - can access actual or usual intake of individuals
        - considered gold standard of dietary intake methods
    - limitations:
        - high respondent burden; participants must be motivated, trained, literate; time-consuming
        - participants often change eating behavior to impress the investigator and/or for ease of recording
- shows an understanding of the limits of nutrient databases and coding
    - database does not contain composition of actual foods; averages of different varieties and samples (how representative is the food they input?)
    - quality of data affected by:
        - skill at estimating portion size
        - forgetting to include some foods/ purposely excluding some foods
    - food substitutions often need to be made (students should outline these)

Analysis/Synthesis/Evaluation

- must provide insightful application of the data
- data and its interpretation are:
    - accurate/correct
    - clearly organized
    - focused; report stays on topic
    - current/academic
    - thorough/complete; shows an understanding of all aspects of the results
- comments are insightful; outlines new/interesting aspects of the information, and applies the information in terms of meaningful suggestions/advice for the participants
- conclusions are based on accurate information and interpretation of the data
- suggestions/advice reflect an insightful understanding and application of the data

Miscellaneous

- report's "production" values show care: organization, writing, proof-reading
- consistent, scholarly style and format used
- referencing is correct and consistent
- includes in-text referencing to give credit to outside sources
- includes short bibliography
- sources are "classics" or recent, say within the past 10 years
- project is handed in on time (or lateness approved by the instructor)
- an absence of "padding," blocks of material known from the course; the project should reflect new or deeper knowledge of nutrition

INFORMED CONSENT FORM

For Participation in diet analysis project for ____.
(name of course)

This diet project is an excellent learning tool in this course, involving analysis, evaluation, synthesis and judgment. The project is optional for those enrolled in the course and serves as an alternate component of the term mark. Participation in the data collection phase is voluntary. It is expected that I will furnish accurate personal data and the details of my diet over three days as best I can to the diet project organizer.

Specifically, I will supply my name, age, weight, height, gender, whether I am pregnant or nursing (if applicable) and my physical activity patterns. I will keep an accurate record of all food consumed over three representative days: the food item's identity, method of preparation, and best estimate of serving size.

The purpose of the study has been clearly explained to me by the student organizer. Typically, this is to compare two diets consumed in differing circumstances by determining the nutrient content of the diets. I will supply general information about my specific circumstances and my food environment - when and where the food was consumed, vegetarian (type), cafeteria meals, prepare all meals, shared cooking, eat out, diabetic diet, food allergies, celebration of birthday/wedding/religious, etc. - as necessary to assist the student to interpret the results.

(If the purpose and circumstances differ significantly from the above examples, I have attached the specifics in my own words and will not partake until this form is countersigned by the course instructor.)

My data will be entered temporarily into a computer program of the organizer, analyzed and be the basis of a report to the course instructor for grading. My identity will be concealed to readers of the report by the author. The report will be returned to the student organizer after grading.

I have had the opportunity to ask questions about the diet project and the data that I am to supply.

I have read and understood the information presented above about the procedures and risks involved in this study and have received satisfactory answers to my questions related to this study. I understand that if I have any questions or concerns resulting from my participation in this study, I may contact:
Name: ______________________________ Telephone no.:_______________________________.
I am aware that I may withdraw from the study at any time, by informing the student investigator in writing. With full knowledge of all foregoing I agree, of my own free will, to participate in this study.

______________________________ ______________________________

Name of Participant/Guardian (Print) Signature

______________________________ ______________________________

Name of Witness (Print) Signature

Date: ____________________ Location: ______________________________

TOPICAL GUIDE TO DIET ANALYSIS MODULES A THROUGH D

The diet analysis modules in this workbook each contain several exercises with worksheets. This guide is provided to allow the reader to quickly identify different exercises pertaining to specific topical areas, and then compare them to choose the individual exercises most appropriate for his or her course.

Topic	Activity(ies) or Exercise(s) Covering Topic for Each Module			
	A	B	C	D
Influences on food choices	1			
Energy nutrient/alcohol intake	3			A, B
Carbohydrates	6	1, 2		A, B
Lipids	7	3, 4		A, B
Protein	8	5	2	A, B
Vitamins	9	6	3, 4	A, B
Minerals	10	7	3, 5	A, B

Modules:

Module A: Group Diet Analysis Activities
Module B: Nutrient Calculations Exercises
Module C: Personal Diet Analysis Exercises
Module D: *Diet Analysis Plus Online* Assignment

Module A—Group Diet Analysis Activities

ACTIVITY DA-1: FOOD BEHAVIOR

se of this exercise is to provide a pleasant beginning of the semester atmosphere and an oductory cooperative learning group activity for students. Students will create a list of factors that influence their food choices.

Instructional Objectives:

Students will

(1) meet and work with two other class members.

(2) become aware of the factors which influence their food choices.

(3) create list of reasons for food choices.

Materials: Assignment sheet: one per group

Time required: 20 minutes

Decisions:

Group size: Three

Assignment to groups: Random assignment (possibly counting off)

Roles: Reader, Recorder, Checker

Reader: Group member who ate something most recently reads the problem out loud, makes certain everyone understands what the group is to do, and encourages all to participate.

Recorder: The group member on reader's left is the recorder who carefully records the best answers of the group members on the group answer sheet, edits what the group has written, gets the group members to check and sign the paper, then turns it in to the instructor.

Checker: The group member on reader's right is the checker who checks on the comprehension or learning of group members by asking them to explain or summarize material learned or discussed. S/he makes sure that everyone understands. S/he sees if all members agree before the recorder records the answer. He lets the group know how much time is left and keeps the group on task so that they finish in the allotted time.

Lesson:

Instructional task:

- Introduce the exercise by stating that the purpose of the exercise is to create a list of factors that affect food choices.
- Form groups of three randomly. Assign roles to group members.
- Review procedure for the cooperative exercise.
  - One list from each group.
  - Make sure everyone participates.
  - Assist everyone in learning.
  - Groups are not competing.

- Hand out the exercise (one sheet per group). Allow 15 minutes for group to create list.

Positive interdependence: One list from each group. Each group member signs final list.

Individual accountability: Randomly select one person from each group to present one reason for food choices to place on list for class.

Expected behaviors: Everyone participates in the discussion and fulfills their role. Each group member can list reasons for food choices.

Monitoring and processing:
Monitoring: Circulate among groups to check that roles are being followed and to answer questions.

Intervening: Remind groups that all members are expected to participate and know about food choice influence.

Processing: Remind groups that every member has two functions: complete the task and work as a group. Ask groups to discuss effectiveness by individually listing things that went well and things that need to be worked on.

DIET ANALYSIS ACTIVITY DA-2: TAGS—IDENTIFYING QUESTIONABLE NUTRITIONAL CLAIMS

Lesson summary:
In this exercise students will look critically at the nutritional claims made in advertising and identify those claims which are questionable.

Instructional Objectives:
The student will
(1) become aware of nutrition claims made in advertising.
(2) identify why a nutrition claim is questionable.
(3) help other group members identify why a nutrition claim is questionable.

Materials: Assignment sheet: one per student

Time required: 20 minutes

Decision:
Group size: Three

Assignment to groups: Random assignment

Roles: Reader, Recorder, Checker.

Reader: Group member with the largest ad reads the problem, makes certain everyone understands what the group is to do, and encourages all to participate.

Recorder: The group member on the reader's left records the tags by putting the checks on the list.

Checker: The group member on the reader's right checks the answer, and keeps the group on task so that they finish in the allotted time.

Roles rotate as each ad is considered.

Lesson:
Instructional task:
- Introduce the exercise by stating the purpose of the exercise.
- Form groups of three randomly. Assign roles.
- Hand out exercise, one sheet per student.
- Review procedure for the exercise.

Positive interdependence: One ad is considered at a time by group.

Expected behaviors: Everyone participates in the discussion and fulfills their roles. Each group member is able to identify questionable nutritional claims.

DIET ANALYSIS ACTIVITY DA-3: PERCENT OF CALORIES EXERCISE

Lesson Summary:
In this exercise students will learn to determine the number of calories derived from classes of nutrients and the percent of calories from each source. They will compare the percent with the Dietary Goals for the United States.

Instructional Objectives:
Students will
(1) calculate number of calories derived from carbohydrate, fat, and protein.
(2) calculate the percent of calories derived from carbohydrate, fat and protein.
(3) compare the percent of calories derived from nutrients with the respective Dietary Goals for the United States.

Materials: Assignment sheet: one per pair during class

Time required: 30 minutes

Decisions:
Group size: Two

Assignment to groups: Random assignment

Roles: Reader, Recorder

Reader: Group member who likes math the most reads the problem, makes certain the other person understands what is to be done.

Recorder: The other group member records the calculations, and keeps the pair on task so that they finish in the allotted time.

Lesson:

Instructional task:

- Introduce the exercise by stating the purpose of the exercise.
- Form groups of two randomly. Assign roles.
- Hand out exercise, one sheet per group.
- Review procedure for the exercise.
- Allow 25 minutes to complete the task.

Positive interdependence: One assignment sheet per pair.

Individual accountability: Randomly select one person from each group to present one reason for food choices to create on list for class.

Expected behaviors: Everyone participates in the discussion and fulfills their role. Each person is able to do the calculations.

DIET ANALYSIS ACTIVITY DA-4: DRI GROUP ACTIVITY

Lesson Summary:

In this exercise students will practice interpreting the Dietary Reference Intake Tables, in particular how to identify their own DRI for a specific essential nutrient. They will also identify which nutrients have RDAs, AIs and ULs, and learn how the Estimated Energy Requirement is determined.

Instructional Objectives:

Students will

(1) identify the categories used in the DRI Tables.

(2) identify the energy nutrients, vitamins and minerals for which there are DRIs, and the minerals for which there are ULs.

(3) describe the method used to determine the Estimated Energy Requirement.

(4) be familiar with the use of the tables to find individual DRIs.

Materials: Assignment sheets: one per student, to be distributed the class period before the group exercise is scheduled; plus, one per group of three during the exercise

Time required: 30 minutes

Decisions:

Group size: Three

Assignment to groups: Random assignment

Roles: Reader/Recorder, Checker, Timer.

Reader/Recorder: Reads the questions to the group and records the group's agreed upon answers on one paper.

Checker: Checks each group member's comprehension status to determine that all group members understand each question.

Timer: Keeps the group on task so as to finish in the allotted time.

Lesson:
Instructional task:
- Distribute the exercise to each student during the class period prior to the exercise.
- Introduce the exercise by stating the purpose of the exercise.
- Form groups of three randomly. Assign roles.
- Hand out exercise, one sheet per group.
- Review procedure for the exercise.
- Allow 30 minutes to complete the task.

Positive interdependence: One assignment sheet per group of three.

Expected behaviors: Everyone prepares their answers to the questions on the handout prior to class and then participates in the discussion, fulfilling their role, during class. Each person is able to use the DRI Tables and identify nutrients included in them.

DIET ANALYSIS ACTIVITY DA-5: FOOD LABEL EXERCISE

Lesson Summary:
In this exercise students will learn how to read food labels. They will read the Nutrition Facts panel and the Ingredients List on a food package. They will learn how this information can be used in selecting food.

Instructional Objectives:
Students will learn
(1) what information is required on a food label.
(2) what information is included in the Nutrition Facts panel of a food label and how it is organized.
(3) what information is provided by a food package Ingredients List and how it is organized.
(4) how this information can be used in a meaningful way in food selection.

Materials: Food packages brought by students
Assignment sheet: one per student

Time required: 20 minutes

Decisions:
Group size: Two

Assignment to groups: Random assignment

Roles: Reader, Recorder

Reader: Group member who brought package reads the problem, makes certain the other person understands what is to be done.

Recorder: The other group member records the information, and keeps the pair on task so that they finish in the allotted time. Roles reverse for second package.

Lesson:

Instructional task:

- Introduce the exercise by stating the purpose of the exercise.
- Form groups of two randomly. Assign roles.
- Hand out exercise, one sheet per student.
- Review procedure for the exercise.
- Allow 10 minutes per package to complete the task.

Positive interdependence: One assignment sheet per pair.

Expected behaviors: Everyone participates in the discussion and fulfills their role. Each person is able to find information on a food package.

DIET ANALYSIS ACTIVITY DA-6: CARBOHYDRATE AND FIBER EXERCISE
DIET ANALYSIS ACTIVITY DA-7: LIPID EXERCISE
DIET ANALYSIS ACTIVITY DA-8: PROTEIN EXERCISE
DIET ANALYSIS ACTIVITY DA-9: VITAMIN EXERCISE
DIET ANALYSIS ACTIVITY DA-10: MINERAL EXERCISE

Lesson Summary:

These five exercises may be completed using either a diet analysis of sample student "Jo Cool" (provided as a handout) or the student's own diet analysis for a single day. As a group, students study the nutrient classes: carbohydrates, lipids, protein, vitamins, and minerals. These exercises ask the student to analyze, evaluate, and make recommendations regarding their own/the sample student's dietary intake. The activity can serve to prepare students for the evaluation of their own three-day diet.

Instructional Objectives:

Students will learn to

(1) identify foods high in specific nutrients.
(2) use guidelines and recommendations to evaluate a menu (or diet).
(3) suggest menu changes that would improve the day's food intake.
(4) calculate percent calories from carbohydrates, lipids, and proteins.

Materials: Assignment sheet for each nutrient class (use Handouts DA-6I to DA-10I if students are using their own 1-day intake reports, or Handouts DA-6S to DA10S if they are using the Jo Cool sample reports); computer analysis for one day's food intake (either the sample analysis or each student's own).

Time required: Varies with exercise (from 20-50 minutes)

by Lorrie Miller Kohler

Decisions:

Group size:	Two or Three
Assignment to groups:	Random assignment
Roles:	Reader, Recorder, Timer

Reader: This group member reads the problem out loud, is responsible for getting information from the computer printout, and makes certain everyone is aware from where the information is obtained.

Recorder: The group member on the reader's right is the recorder who is responsible for writing the information from the computer printout. S/he also checks on the comprehension or learning of group members by asking them to explain or summarize material learned or discussed. S/he makes sure that everyone understands.

Timer: The timer is at the reader's left. S/he lets the group know how much time is left and keeps the group on task so that they finish in the allotted time. S/he is responsible for handing in any work from the group.

Lesson:

Instructional task:

- Introduce the exercise by stating what the purpose of the exercise is.
- Form groups of two or three. Assign roles to group members.
- Review procedure for the cooperative exercise.
  - Make sure everyone participates.
  - Assist everyone in learning.
  - Groups are not competing.
- Hand out exercise, one sheet per person.
- Announce amount of time allotted.

Positive interdependence: A mixture of individual and group tasks. Only one computer data sheet per group (if the provided sample student printout is used).

Expected behaviors: Everyone participates in the discussion and fulfills their role. There will be a sharing of ideas and learning from each other.

Monitoring and processing:

Monitoring: Circulate among groups to check that roles are being followed and to answer questions.

Intervening: Remind groups that all members are expected to participate and to help each other.

Processing: Remind groups that every member has two functions: complete the task and work as a group. Ask groups to discuss effectiveness by individually listing things that went well and things that need to be worked on.

by Lorrie Miller Kohler

Handout DA-1 (page 1 of 2)

Group members' signatures:

____

____

____

DIET ANALYSIS ACTIVITY DA-1: FOOD BEHAVIOR

Objectives:

- To meet and work with two other class members
- To become aware of the factors which influence food choices
- To create a list of reasons for food choices

Instructions:

1. Individually (on another sheet of paper) write down 3 of your most favorite foods and 3 foods you intensely dislike. (Take only 2 minutes.)

You have fifteen minutes to complete the remainder of this activity.

2. In your group, compare your likes and dislikes, and determine what caused your acceptance/non-acceptance of these foods.

3. Your group will make a list of the **reasons** for which you made your food choices. Choose a recorder for your group. The recorder will record the **reasons** for the likes and dislikes in Table I found back side of this sheet.

4. After reasons for the likes and dislikes have been recorded, each person sign off on the front side of this sheet. The recorder hands in this sheet to the instructor.

Table I: Reasons for Food Choices: Likes and Dislikes

Person	**Reason for Likes**	**Reason for Dislikes**
Person:	1. 2. 3.	1. 2. 3.
Person:	1. 2. 3.	1. 2. 3.
Person:	1. 2. 3.	1. 2. 3.

***After recording data, all group members remember to sign off on front of sheet**

Name______________________________

Diet Analysis Activity DA-2: Tags—Identifying Questionable Nutritional Claims

Objectives:

- To look critically at the nutritional claims made in advertising.
- To become able to identify why a nutrition claim is questionable.
- To help other group members identify why a nutrition claim is questionable.

Instructions:

Each student was instructed to bring to class today an advertisement which made a nutrition claim, or a written statement of an ad from radio or television.

1. Each of you identify and briefly write the nutritional information given in your ad.

2. Write a statement about how accurate you believe this information is and defend your statement.

3. In groups of three, look at the ads brought by the group members, one at a time. For each, determine and agree which of the tags in the list below could be tied to each ad by checking it off on the list. The more tags you can tie on the ad, the less likely the claims are valid. Use no more than 5 minutes per ad.

TAGS

☐ Logic without proof	☐ Misuse of terms
☐ Amateur diagnosis	☐ Scare tactics
☐ Sales pitch	☐ Motive: Personal gain
☐ Magical thinking	☐ Bent truth
☐ Authority not cited	☐ Unreliable publication
☐ Fake credentials	☐ Incomplete truth

☐ Logic without proof	☐ Misuse of terms
☐ Amateur diagnosis	☐ Scare tactics
☐ Sales pitch	☐ Motive: Personal gain
☐ Magical thinking	☐ Bent truth
☐ Authority not cited	☐ Unreliable publication
☐ Fake credentials	☐ Incomplete truth

TAGS

☐ Logic without proof	☐ Misuse of terms
☐ Amateur diagnosis	☐ Scare tactics
☐ Sales pitch	☐ Motive: Personal gain
☐ Magical thinking	☐ Bent truth
☐ Authority not cited	☐ Unreliable publication
☐ Fake credentials	☐ Incomplete truth

4. Each person staple his ad to his assignment sheet. As a group determine which ad made the most unfounded claims. Staple all sheets of the group together with this one on the top. Hand in.

Extension exercise: If you finish before the time is up, join another group to see what nutritional claims were made in the ads they brought.

Name___________________

Name___________________

DIET ANALYSIS ACTIVITY DA-3: PERCENT OF CALORIES EXERCISE

Objectives:

- To calculate number of calories derived from carbohydrate, fat, and protein.
- To calculate the percent of calories derived from carbohydrate, fat and protein.
- To compare the percent of calories derived from nutrients with the respective Dietary Goals for the United States.

Instructions:

The Nutrition Facts on a box of granola cereal shows that a 1-oz. serving (with no milk) provides the following:

calories	110
protein	2 grams
carbohydrate	22 grams
fat	2 grams

Answer the following questions. Use a calculator to help, but show what calculations you have done.

You will work in pairs. Be certain that both of you understand what to do and participate. Be certain that the exercise is done correctly. Watch the time. Each group hands in only one set of answers. Be sure that both names are on the answer sheet. Be sure both group members understand how to do these calculations and comparisons. Be prepared to discuss the questions.

1. How many calories in this cereal come from:

   a. carbohydrate? b. fat? c. protein?

2. Using 110 as the total number of calories in a 1-oz. serving, calculate what percent of calories in this cereal come from:

   a. carbohydrate   b. fat   c. protein

3. How does the percent of calories from each nutrient calculated in Question 2 compare with the U.S. Dietary Goals for:

   a. carbohydrate?

   b. fat?

   c. protein?

by Lorrie Miller Kohler

Names:_______________________

____

____

DIET ANALYSIS ACTIVITY DA-4: DRI GROUP ACTIVITY

Directions:
Each person answer the following questions about the Dietary Reference Intakes (DRI) on this sheet. Refer to your textbook for information on essential nutrients and nutrient recommendations. You will receive 5 points for answering these questions before the next class meeting. At the next class meeting, you will work in groups of 3 to share your answers and to determine the best answer to each of these questions. At that time, your instructor will provide each group a single copy of this assignment sheet. One person in the group will record the group's agreed upon answers on the assignment sheet. You will have 30 minutes to complete your answers. Each group is responsible for turning in <u>one</u> paper with the group's agreed upon answers and <u>three</u> signatures on it. Your signature on the paper will earn you an additional 5 points for participating in the group process. Staple your completed individual assignments to the single assignment sheet completed by the group; turn all in at the end of the class.

Objectives:
Students will learn:

- how to interpret the Dietary Reference Intake Tables including categories and essential nutrients found there.
- what essential nutrients have RDAs, AIs, ULs.
- how to determine what student's specific DRI is for a particular essential nutrient.
- how the Estimated Energy Requirement is determined.

Designate the following roles in the group:

Reader/Recorder: Reads the questions to the group and records the group's agreed upon answers on one paper.

Checker: Checks each group member's comprehension status to determine that all group members understand each question.

Timer: Keeps the group on task so as to finish in the allotted time.

1. Determine the **categories** (e.g., Infants) one should look at when trying to determine the recommended nutrient intake for a particular nutrient for a given individual. List them here.

2. List the **energy nutrients** for which there is an RDA.

3. List the **fat-soluble vitamins** for which there is a DRI (RDA or AI).

4. List the **water-soluble vitamins** for which there is a DRI (RDA or AI).

5. List the **major minerals** for which there is a DRI (RDA or AI).

6. What is done with the **major minerals** for which there is no DRI (RDA or AI)? Which are they?

7. List the **trace minerals** for which there is a DRI (RDA or AI).

8. For which **major minerals** and **trace minerals** is there a **UL**? List them below.

9. Describe how the **Estimated Energy Requirement** for energy is determined.

10. Each person in the group select an essential nutrient and determine <u>your</u> recommended intake for the nutrient. In the space below, list the nutrients and the DRI (RDA or AI) for the essential nutrient each group member has chosen.

	Essential Nutrient	DRI (RDA or AI)
1.		
2.		
3.		

by Lorrie Mill

Handout DA-5

Name of person who brought food label__________________
Name of second group member __________________

DIET ANALYSIS ACTIVITY DA-5: FOOD LABEL EXERCISE

Objectives:

- To read food labels.
- To learn what information is required on a food label.
- To learn what information is included in the Nutrition Facts panel of a food label and how it is organized.
- To learn what information is provided by a food package Ingredients List and how it is organized.
- To learn how this information can be used in a meaningful way in food selection.

Instructions:

Each student is to bring a food package to class. The package should have both an Ingredients List and a Nutrition Facts panel on it. Work in groups of two.

Consider one package. The person who brought the package will suggest the answers to the questions. The pair will discuss the answers and decide on what the second person will write on the assignment sheet. Hand in the sheet for the food package with the package (or the appropriate part of it) attached. Fill out the assignment sheet for the second package in a similar way.

Food Label: Items 1, 2, and 3 are three of the items required by law to appear on a food label. List them for your food package.

1. The common name of the product. ____________________________________

2. The name and address of the manufacturer, packer, or distributor.

3. The net contents in terms of weight, measure, or count. _________________________

4. What nutrient content claims, if any, are made on the front of the package?

5. What health claims, if any, are made regarding the product?

Ingredient List

6. List the ingredient that is present in the greatest proportion by weight.

7. List the ingredient that is present in the second greatest proportion by weight.

Nutrition Facts panel

8. List the following nutrition information for the product:
   Serving size: ____________
   Servings per container: ____________
   Calories per serving: ____________

Nutrition Facts	**Amount per serving**	**% Daily Value***
Total fat(g)	____________	____________
Saturated Fat(g)	____________	____________
Cholesterol(mg)	____________	____________
Sodium(mg)	____________	____________
Total carbohydrate(g)	____________	____________
Dietary fiber(g)	____________	____________
Sugars(g)	____________	____________
Protein(g)	____________	____________
Vitamin A	____________	____________
Vitamin C	____________	____________
Calcium	____________	____________
Iron	____________	____________

*Based on a 2000 calorie diet

9. List the % Daily Values that appear to be high or low for this product.

10. In what ways can the information from a food label help you eat a better diet?

FORMAL COOPERATIVE LEARNING GROUP ACTIVITIES

This semester you will have the opportunity to work in Formal Cooperative Learning Groups. In these groups you will work with a group of people in what is called a base group, have a specific role in the group, and follow a particular group process. The ultimate goal of formal cooperative group learning and process is to ensure the learning of each member of the group.

Early on in the semester each student will be assigned to a cooperative learning base group with three other students with whom they will work for the remainder of the semester. The task of the base group is to discuss the group activities 6S-10S. A simulated printout of a one-day dietary intake for student "Jo Cool" is to be used as the basis for answering questions found in the activities on carbohydrates, lipids, proteins, vitamins and minerals. Your instructor will provide due dates for when the activities are to be completed in class. On those dates, the questions in each activity will be discussed in the assigned base groups. Since the intent of cooperative group learning is to foster the group's learning, it is important that all members take seriously their responsibility to complete the activities. After the group discussion is finished, each group will complete a short written evaluation on how their group functioned. Then, as a class, we will summarize the answers to the questions that comprise the activity. Since the intent of the group process is that all members of the group learn the content of a particular activity, all should be prepared to answer any question if asked to do so. All group members will turn in the particular group activity and their group's evaluation form in the assigned group's folder after the discussion is completed.

Guidelines for Cooperative Groups:
The following guidelines are important to follow when working in cooperative learning groups:

1. Learn the names of your group members.
2. When your group meets, arrange chairs so you are facing each other; everyone should be able to hear and see all members and feel included.
3. Each person should have a role in the group. Determine roles before you begin the discussion of the activity. Rotate roles at each group meeting.
4. Each student is responsible to him/herself and to his/her team.
5. Each student has an obligation to learn the material and to try to ensure that all teammates learn it.

Social Skills for Cooperative Groups:
The following social skills are important to effectively facilitate the cooperative group process:

1. Tolerant Listening
2. Constructive Disagreement
3. Asking for Clarification
4. Expressing Need for Assistance
5. Summarizing

Roles and Tasks:

The following roles and tasks will be used in our cooperative groups:

1. Reader: reads aloud the questions in the exercise to the group members.
2. Checker: asks each person to state their answer to the questions and asks if all agree/disagree. All group members should agree on answer to question and understand answer.
3. Summarizer: summarizes the answer to each question after discussion of the question is completed by the group.
4. Timer: watches the time, keeps group on task and makes sure group finishes in allotted time. At the completion of the exercise, timer sees that all papers are placed in the group's folder and that ABSENTEES are recorded on the attendance sheet inside the folder. Folders are turned in to instructor.

by Lorrie Miller Kohler

Name______________________

DIET ANALYSIS ACTIVITY DA-6S: CARBOHYDRATE AND FIBER EXERCISE

Objectives:

- To identify foods high in simple and complex carbohydrates and in fiber.
- To use guidelines and recommendations to evaluate a menu (or diet).
- To suggest menu changes that would improve carbohydrate and fiber in the diet.
- To calculate percent calories from carbohydrates.

Instructions:
Answer the following questions about carbohydrates, fiber and calories using the simulated computer printout for Jo Cool (pp. 111-117).

1. Using the Intake Spreadsheet, list the foods eaten by Jo Cool in order from highest to lowest according to amount (in grams) of carbohydrate they contain. Also record the quantity of food. Do the same for fiber-containing foods.

Carbohydrate-rich foods (more than 5 grams)		**Fiber-rich foods (more than 1 gram)**	

2. Which foods contain little or no carbohydrate (less than 1 gram)? fiber (less than 1 gram)?

Little/no carbohydrate	**Little/no fiber**		

3. a. Are the foods listed in #1 which are high in carbohydrate and fiber of plant or animal origin?

   b. Are the foods listed in #2 of plant or animal origin?

   c. From questions 3a and b, what do you conclude about the sources carbohydrates and fiber?

4. Which foods contain complex carbohydrates (starch)? simple carbohydrates?

Complex carbohydrates (starch)	Simple carbohydrates

5. **Each group member** suggest a change you would make to increase the complex carbohydrates (starch) in this day's intake. Record suggestions below.

6. Would you make any changes in the fiber intake? If so, describe one. If not, why not? Record group members' changes or reasons for not changing below.

7. Calculate the percent of calories in the doughnut that came from carbohydrate. Be sure **all** group members understand how to do these calculations. Show your math.

8. Calculate what percent of the total calories came from carbohydrate. Be sure **all** group members understand how to do these calculations. Show your math.

9. In a sentence, describe how the percent of total calories from carbohydrate in this day's intake compares with the Dietary Goals.

10. Do you think this day's intake implemented the *Dietary Guidelines for Americans* related to calories, carbohydrates and fiber? (Review *Dietary Guidelines*.) If so, how? If not, how not? Record group members' responses below.

Computer Printout for Jo Cool (page 1 of 7)

Profile

Profile Name	Jo Cool
Height	5 ft. 6 inches
Weight	130 lb.
Age	23 years
BMI	21

DRI Goals

Nutrient	DRI	
Energy Nutrients		
Calories (kcal)	2176 kcal	
Carbohydrates	245 - 354 g	*45%-65% of kilocalories*
Fat	48 - 85 g	*20%-35% of kilocalories*
Protein	54 - 190 g	*10%-35% of kilocalories*
Protein (g/kg)	47 g	*Daily requirement based on grams per kilogram of body weight*
Fat		
Saturated Fat	24	*less than 10% of calories recommended*
Monounsaturated Fat	-	*No recommendation*
Polyunsaturated Fat	-	*No recommendation*
Cholesterol	300 mg	*less than 300mg recommended*
Essential fatty acids		
PFA 18:2, Linoleic	12 g	
PFA 18:3, Linolenic	1.1 g	
carbohydrates		
Dietary Fiber, Total	25 g	
Sugar, Total	-	*No recommendation*
Other		
Water	2.3 L	
Alcohol	-	*No recommendation*
Vitamins		
Thiamin (Vit B1)	1.1 mg	
Riboflavin	1.1 mg	
Niacin	14 mg	
Vitamin B6	1.3 IU	
Vitamin B12	2.4 µg	
Folate (DFE)	400 µg	
Vitamin C	75 mg	
Vitamin D (ug)	5 µg	
Vitamin A(RAE)	699.9 µg	
alpha-tocopherol (Vit E)	15 mg	
Minerals		
Calcium	1000 mg	*DRI Adequate Intake*
Iron	18 mg	
Magnesium	310 mg	
Potassium	4700 mg	*DRI Adequate Intake*
Zinc	8 mg	
Sodium	1500 mg	*DRI Adequate Intake*

!	Nutrient	Intake	DRI	0% 50% 100%
	Energy			
	Calories	2177 kCal	2176.3 kCal	100%
	Carbohydrates	231 g	245 g - 354 g	
	Fat	97 g	48 g - 85 g	
	Protein	104 g	54 g - 190 g	
	Protein(g/kg/day)	104 g	47 g	221%
	Fat			
	Saturated Fat	30.81 g	*no rec*	
	Monounsaturated Fat	36.05 g	*no rec*	
	Polyunsaturated Fat	17.39 g	*no rec*	
	Cholesterol	284.63 mg	300 mg	95%
	Essential fatty acids			
	PFA 18:2, Linoleic	15.33 g	12 g	128%
	PFA 18:3, Linolenic	1.46 g	1.1 g	133%
	carbohydrates			
	Dietary Fiber, Total	13.17 g	25 g	53%
	Sugar, Total	88.7 g	*no rec*	
	Other			
	Water	1.54 L	2.3 L	67%
	Alcohol	0 g	*no rec*	
	Vitamins			
	Thiamin (Vit B1)	2.44 mg	1.1 mg	222%
	Riboflavin	2.97 mg	1.1 mg	270%
	Niacin	28.58 mg	14 mg	204%
	Vitamin B6	2.4 IU	1.3 IU	185%
	Vitamin B12	9.08 µg	2.4 µg	378%
	Folate (DFE)	458.26 µg *	400 µg	115%
	Vitamin C	91.01 mg	75 mg	121%
	Vitamin D (ug)	5.02 µg *	5 µg	100%
	Vitamin A(RAE)	739.78 µg	699.9 µg	106%
	alpha-tocopherol (Vit E)	5.56 mg *	15.0 mg	37%
	Minerals			
	Calcium	839.91 mg	1000 mg	84%
	Iron	19.77 mg	18 mg	110%
	Magnesium	300.58 mg	310 mg	97%
	Potassium	3229.88 mg	4700 mg	69%
	Zinc	16.08 mg	8 mg	201%
!	Sodium	2487.32 mg	1500 mg	166%

Item Name	*Quantity*	*Weight*	Kcal (kcal)	Protein (g)
Cake Doughnut	1.0 item	47 g	197.86	2.34
Mashed Potatoes w...	1.0 c.	208.96 g	173.43	4.1
Mayonnaise with S...	1.0 t.	13.80 g	98.94	0.15
Butter	1.0 t.	15 g	107.55	0.12
Brewed Coffee	2.0 c.	474 g	18.95	0.66
Orange Juice, Uns...	0.50 c.	124.50 g	56.2	0.84
Peanut Butter, Sm...	1.50 t.	24 g	143.75	5.99
Catsup or Ketchup	1.0 t.	15 g	14.25	0.27
Plain Hamburger R...	1.0 item	43 g	119.97	4.8
Nonfat, Skim or F...	0.50 c.	122.50 g	41.65	4.12
Pork Loin, Center...	3.0 oz.	0 g	197.19	27.35
Cabbage, Shredded	0.50 c.	35 g	8.40	0.50
Yellow Sweet Corn...	0.50 c.	105 g	82.94	2.53
Chocolate Pudding...	1.0 c.	284 g	337.95	9.11
Ground Beef, Regu...	4.0 oz.	0 g	330.93	30.82
Nonfat, Skim or F...	1.0 c.	245 g	83.30	8.25
RITZ Crackers	4.0 item	12.80 g	64	0.80
KELLOGG'S CORN FL...	1.0 c.	28 g	100	2
Totals:			**2177.26 kcal**	**104.75 g**

Item Name	Carb (g)	Fat (g)	Sat Fat (g)	Mono Fat (g)
Cake Doughnut	23.35	10.76	1.70	4.36
Mashed Potatoes w...	36.62	1.19	0.63	0.25
Mayonnaise with S...	0.53	10.79	1.64	2.70
Butter	0	12.16	6.12	5
Brewed Coffee	0	3.60	0	0
Orange Juice, Uns...	13.42	0.7	0	0.1
Peanut Butter, Sm...	4.42	12.54	2.40	5.93
Catsup or Ketchup	3.58	0.8	0.1	0.1
Plain Hamburger R...	21.26	1.86	0.46	0.47
Nonfat, Skim or F...	6.7	0.9	0.14	0.5
Pork Loin, Center...	0	8.89	3.9	3.78
Cabbage, Shredded	1.95	0.4	0	0
Yellow Sweet Corn...	20.41	0.52	0.8	0.15
Chocolate Pudding...	55.9	8.94	5.13	2.31
Ground Beef, Regu...	0	22.5	8.67	9.65
Nonfat, Skim or F...	12.15	0.19	0.28	0.11
RITZ Crackers	8	3.20	0.40	1.20
KELLOGG'S CORN FL...	24	0	0	0
Totals:	**232.29 g**	**99.94 g**	**32.37 g**	**36.61 g**

Item Name	Poly Fat (g)	Chol (mg)	Linoleic (g)	Linolenic (g)
Cake Doughnut	3.70	17.38	3.42	0.25
Mashed Potatoes w...	0.13	4.17	0.9	0.4
Mayonnaise with S...	5.88	5.24	5.19	0.68
Butter	0.43	32.25	0.6	0
Brewed Coffee	0	0	0	0
Orange Juice, Uns...	0.1	0	0.1	0
Peanut Butter, Sm...	3.57	0	3.55	0.1
Catsup or Ketchup	0.3	0	0.3	0
Plain Hamburger R...	0.84	0	0.75	0.8
Nonfat, Skim or F...	0	2.45	0	0
Pork Loin, Center...	1.13	78.19	0.99	0.7
Cabbage, Shredded	0.2	0	0	0.1
Yellow Sweet Corn...	0.24	0	0.24	0
Chocolate Pudding...	0.50	25.55	0.32	0.18
Ground Beef, Regu...	0.82	114.46	0.61	0.9
Nonfat, Skim or F...	0.1	4.90	0	0
RITZ Crackers	0	0	0	0
KELLOGG'S CORN FL...	0	0	0	0
Totals:	**17.94 g**	**284.59 mg**	**16.97 g**	**4.11 g**

Item Name	Diet Fiber (g)	Sugar (g)	Water (L)	Alcohol (g)
Cake Doughnut	0.70	7.3	0	0
Mashed Potatoes w...	3.13	3.15	0.16	0
Mayonnaise with S...	0	0.6	0	0
Butter	0	0	0	0
Brewed Coffee	0	0	0.47	0
Orange Juice, Uns...	0.24	10.45	0.10	0
Peanut Butter, Sm...	1.41	1.87	0	0
Catsup or Ketchup	0.19	3.7	0.1	0
Plain Hamburger R...	0.90	2.68	0.1	0
Nonfat, Skim or F...	0	6.23	0.11	0
Pork Loin, Center...	0	0	0.4	0
Cabbage, Shredded	0.80	1.25	0.3	0
Yellow Sweet Corn...	2.9	3.73	0.8	0
Chocolate Pudding...	2.27	33.85	0.20	0
Ground Beef, Regu...	0	0	0.5	0
Nonfat, Skim or F...	0	12.47	0.22	0
RITZ Crackers	0.40	0.80	0	0
KELLOGG'S CORN FL...	1	2	0	0
Totals:	**13.94 g**	**90.8 g**	**3.46 L**	**0 g**

Item Name	Vit B1 (mg)	Ribo (mg)	Niacin (mg)	Vit B6 (IU)
Cake Doughnut	0.10	0.11	0.87	0.2
Mashed Potatoes w...	0.18	0.8	2.34	0.48
Mayonnaise with S...	0	0	0	0.7
Butter	0	0	0	0
Brewed Coffee	0	0.23	0	0
Orange Juice, Uns...	0.9	0.2	0.25	0.5
Peanut Butter, Sm...	0.1	0.2	3.21	0.10
Catsup or Ketchup	0	0.7	0.22	0.2
Plain Hamburger R...	0.17	0.13	1.78	0.3
Nonfat, Skim or F...	0.5	0.22	0.11	0.4
Pork Loin, Center...	1.5	0.27	5.9	0.43
Cabbage, Shredded	0.1	0.1	0.10	0.3
Yellow Sweet Corn...	0.4	0.7	1.22	0.5
Chocolate Pudding...	0.10	0.54	0.42	0.8
Ground Beef, Regu...	0.4	0.23	7.33	0.33
Nonfat, Skim or F...	0.11	0.44	0.23	0.9
RITZ Crackers	0.5	0.3	0.35	0
KELLOGG'S CORN FL...	0.37	0.42	5	0.50
Totals:	**5.43 mg**	**5.59 mg**	**29.33 mg**	**6.64 IU**

Item Name	Vit B12 (µg)	Fol (DFE) (µg)	Vit C (mg)	Vit D (ug) (µg)
Cake Doughnut	0.12	-	0.9	-
Mashed Potatoes w...	0.14	16.71	12.95	0
Mayonnaise with S...	0.3	1.10	0	0.13
Butter	0.2	0.44	0	0.20
Brewed Coffee	0	9.47	0	0
Orange Juice, Uns...	0	54.77	48.43	0
Peanut Butter, Sm...	0	17.76	0	0
Catsup or Ketchup	0	2.25	2.26	0
Plain Hamburger R...	0.8	73.9	0	-
Nonfat, Skim or F...	0.64	6.12	0	1.22
Pork Loin, Center...	0.64	5.9	1.1	-
Cabbage, Shredded	0	15.5	11.26	0
Yellow Sweet Corn...	0	51.45	8.50	0
Chocolate Pudding...	0.85	11.35	0	-
Ground Beef, Regu...	3.71	11.33	0	0
Nonfat, Skim or F...	1.29	12.25	0	2.45
RITZ Crackers	0	-	0.48	-
KELLOGG'S CORN FL...	1.50	169.96	6	1
Totals:	**10.19 µg**	**460.26 µg**	**91.88 mg**	**5 µg**

Item Name	Vit A (RAE) (µg)	alpha-T (mg)	Calcium (mg)	Iron (mg)
Cake Doughnut	8.11	1.44	20.68	0.91
Mashed Potatoes w...	11.39	0.8	45.97	0.56
Mayonnaise with S...	11.70	0.25	2.48	0.6
Butter	113.57	0.18	3.59	0
Brewed Coffee	0	0	4.73	0.4
Orange Juice, Uns...	40.36	0.18	11.20	0.12
Peanut Butter, Sm...	0	1.92	11.27	0.45
Catsup or Ketchup	42.40	0.17	2.70	0.7
Plain Hamburger R...	0	0.53	59.34	1.42
Nonfat, Skim or F...	75.71	0.3	111.47	0.61
Pork Loin, Center...	1.80	0.17	19.54	0.83
Cabbage, Shredded	18.13	0.2	16.45	0.20
Yellow Sweet Corn...	25.77	0.7	5.25	0.44
Chocolate Pudding...	75.72	0.13	272.64	0.99
Ground Beef, Regu...	0	0.20	13.59	3.10
Nonfat, Skim or F...	151.43	0.7	222.94	1.22
RITZ Crackers	12.12	-	16	0.57
KELLOGG'S CORN FL...	151.50	0.2	0	8.10
Totals:	**739.71 µg**	**8.7 mg**	**839.84 mg**	**21.22 mg**

Item Name	Magn (mg)	Potas (mg)	Zinc (mg)	Sodium (mg)
Cake Doughnut	9.39	59.68	0.25	256.61
Mashed Potatoes w...	37.61	618.50	0.60	631.4
Mayonnaise with S...	0.13	4.69	0.2	78.38
Butter	0.30	3.59	0.1	86.40
Brewed Coffee	9.47	227.52	0.4	4.73
Orange Juice, Uns...	12.44	236.55	0.6	1.24
Peanut Butter, Sm...	42	132.47	0.70	120
Catsup or Ketchup	2.84	57.29	0.3	166.94
Plain Hamburger R...	9.2	40.41	0.28	205.97
Nonfat, Skim or F...	11.2	118.82	1.4	53.90
Pork Loin, Center...	27.20	381.64	2.7	73.9
Cabbage, Shredded	5.25	86.10	0.6	6.30
Yellow Sweet Corn...	24.14	195.30	0.48	3.15
Chocolate Pudding...	56.79	426	1.36	278.32
Ground Beef, Regu...	24.93	370.60	6.58	105.40
Nonfat, Skim or F...	22.4	237.64	2.8	107.80
RITZ Crackers	2.56	8	0.18	108
KELLOGG'S CORN FL...	3.35	25	0.16	200
Totals:	**301.20 mg**	**3229.80 mg**	**19.69 mg**	**2488.44 mg**

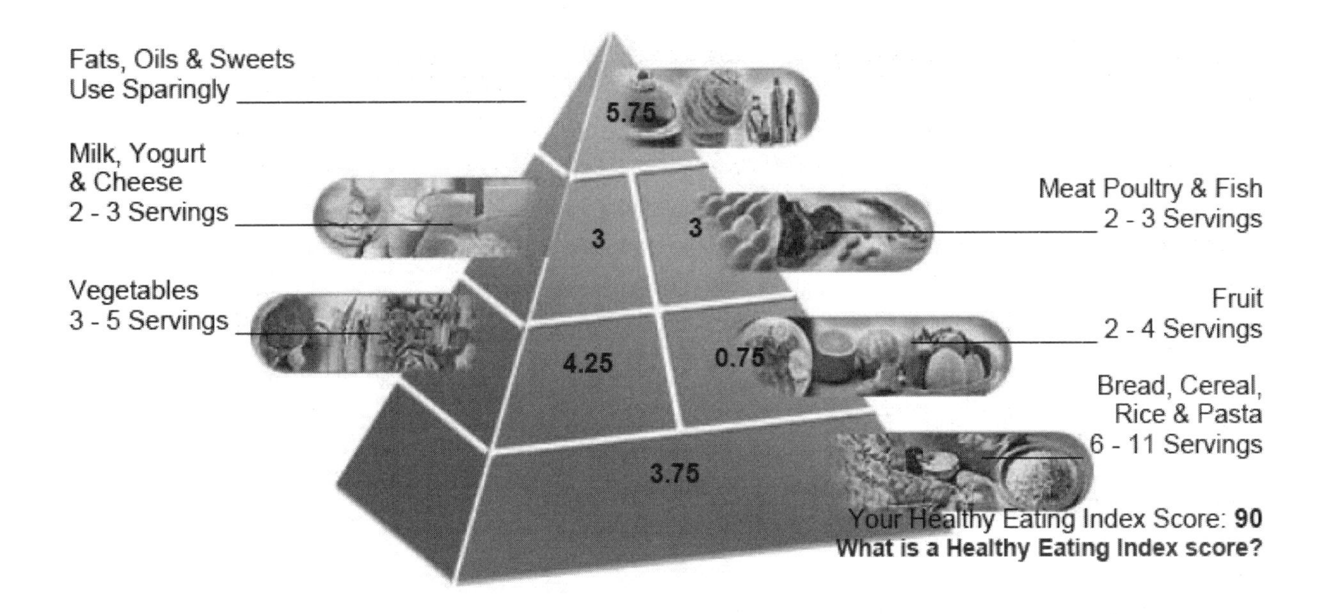

Macronutrient Ranges

	Recommended		Yours		In range?
Carbohydrates	45%-65%	979-1414 kCal	42%	909 kCal	No
Fats	20%-35%	435-762 kCal	39%	859 kCal	No
Proteins	10%-35%	218-762 kCal	19%	409 kCal	Yes

Fat as Percentage of Total Calories

Source of Fat	0% – 25% – 50% – 75% – 100%
Saturated Fat	12.73%
Mono Fat	14.9%
Poly Fat	7.19%
Cholesterol	0.11%
Other/Unspecified	0%

Name________________________

DIET ANALYSIS ACTIVITY DA-7S: LIPID EXERCISE

Objectives:

- To identify foods high in saturated and unsaturated fats, and in cholesterol.
- To calculate percent fat by weight and percent calories from fat.
- To use guidelines and recommendations to evaluate a menu for fat content.
- To suggest dietary changes that would improve fat and cholesterol intake.

Instructions:

Answer the following questions about fats, cholesterol, and calories using the simulated computer printout for Jo Cool (pp. 111-117).

1. Review the food Intake Spreadsheet from the simulated computer printout. Identify and list foods you believe are high in total fat and in cholesterol.

Foods high in total fat	Foods high in cholesterol

2. a. Using the Intake Spreadsheet, identify those foods containing 3 or more grams of total fat, saturated fat, monounsaturated fat, and polyunsaturated fat, and list them in order from high to low. Also record quantity of food.

Total fat	Saturated fat	Monounsat. fat	Polyunsat. fat

b. Do the same for those foods containing 5 or more milligrams of cholesterol. Record milligrams of cholesterol and quantity of food.

Cholesterol			

3. The ground beef patty weighs 113 grams, 22.5 grams of which are fat. Calculate % fat by weight in this ground beef patty by dividing the weight of the fat by the total weight of the patty and multiply by 100.

   22.5 grams/113 grams =_____ x 100 =______% fat by weight in the ground beef patty

4. Calculate % calories from fat for the ground beef patty.
   Step 1. Calculate the number of calories in 22.5 grams of fat.
   Step 2. Divide that number by total calories in the patty (330.93 kcal).
   Step 3. Multiply by 100.

   Calories from fat/total patty cal. =____ x 100% =____ % cal. from fat in ground beef patty

5. Write a single sentence describing how the answer from Question #3 is different from the answer from Question #4.

6. If the total caloric intake for this day is 2177.26 calories, then what percent of total calories was contributed by total fat in the beef patty?
   Calculate this by:
   Step 1. Multiplying 22.5 grams x 9 = calories from fat in beef patty.
   Step 2. Dividing this number by 2177.26 and multiplying by 100.

   Calories from fat/total calories for day x 100 = ____% calories from fat in the beef patty

7. Note how Questions 3, 4, and 6 are asking different questions. As a group agree on and summarize the differences in a single sentence and write it below.

8. What percent of the total calories come from saturated, monounsaturated, and polyunsaturated fat? The total grams of these fats is given in the Intake Spreadsheet. To calculate the amounts, multiply the number of grams of each type of fat by 9; then divide each answer by 2177.26 (total calories); and multiply by 100. Show your math.

_______% calories from saturated fat

_______% calories from monounsaturated fat

_______% calories from polyunsaturated fat

9. How do the values in Question 8 compare with recommendations for total fat intake of 20%-35% of calories with less than 10% of calories coming from saturated fat? As a group summarize the comparisons in a sentence and write it here.

10. Would you recommend any changes in the foods containing fats that were selected in this day's intake? If yes, what suggestion(s)? If not, why not? Record group's responses below.

11. In Question #2 you listed foods containing cholesterol from in this day's intake. What is the origin—plant or animal—of these foods?

12. What is the source of cholesterol in the doughnut: __________, mashed potatoes:__________, mayonnaise:__________, and pudding:__________?

13. How does the total cholesterol consumed in this day's intake compare with the recommendation to consume less than 300 milligrams of cholesterol a day?

14. Would you make any suggestions for change in cholesterol intake of Jo Cool? If yes, what suggestion(s)? If no, why not? Record changes or reasons for no change below.

by Lorrie Miller Kohler

Name______________________

Diet Analysis Activity DA-8S: Protein Exercise

Objectives:

- To identify foods high in quantity of protein.
- To evaluate the adequacy of protein intake in grams and in the context of caloric intake.
- To compare the quality of protein in foods from plant and animal sources.
- To calculate percent calories from protein.
- To suggest menu choices that are high in protein and fiber and low in fat.

Instructions:
Answer the following questions about proteins using the simulated computer printout for Jo Cool (pp. 111-117).

1. Review the foods eaten by Jo Cool using the Intake Spreadsheet and identify the foods that contain more than 3 grams of protein. List these foods in decreasing order of amount of protein and according to their source. Also record quantity of food.

Animal-derived protein		**Plant-derived protein**	

2. Compare the quantity of protein obtained from these two sources. As a group agree on and describe in a single sentence the comparison. Record it below.

3. Describe and compare the total number of grams of protein (_____) with the RDA for protein (_____) for Jo Cool. * Remember the number of grams of protein consumed should be no more than twice the RDA for protein. What is twice the RDA for Jo Cool?_____ Make this comparison in the context of looking at her calorie intake (______) for the day and her DRI for calories (______). In looking at the gram intake in the context of the caloric intake, is the quantity of her protein intake for this day adequate? Why or why not?

4. Each group **member** compare the quality of protein from the two sources (in Question #1). As a group describe the comparison in a single sentence. Record it below.

5. In the space below rank all the protein-rich foods listed in Question #1 from highest to lowest according to the amount (in grams) of total fat. Record number of grams of fat.

6. After looking at the data in Question #5, as a group, make a statement comparing the source of high protein-rich foods with their fat content.

7. What percent of the total calories for the day came from protein? Show your calculations below.

8. Do you think this day's intake implemented the DRI goals for protein? If yes, how? If no, how not? Record group members' responses.

9. As a group delete one of this day's high-protein foods and add another protein source which **would not change the amount of protein, but would decrease the amount of total fat and increase the amount of fiber.** Use a food composition table to verify that your choice is a good one.

Name the food you are deleting. ______________________________

- How much protein does it contain? ________
- How much fat does it contain? ________
- How much fiber does it contain? _______

Name and give the amount of the food you are adding. _________________________

- Indicate the amount of protein in the food you are adding ________
- Indicate the amount of fat in the food you are adding __________
- Indicate the amount of fiber in the food you are adding __________

Name______________________________

DIET ANALYSIS ACTIVITY DA-9S: VITAMIN EXERCISE

Objectives:

- To identify food sources of selected vitamins.
- To use RDAs to evaluate intake of selected vitamins.
- To suggest menu choices that would provide adequate intake of selected vitamins.
- To identify major functions of selected vitamins.

Instructions:
Answer the following questions about vitamins using the simulated computer printout for Jo Cool (pp. 111-117).

1. Review the foods eaten by Jo Cool using the Intake Spreadsheet and identify the foods that contain 10 milligrams or more of vitamin C. List these foods in <u>decreasing order</u> of amount in milligrams of vitamin C they contain. Also record quantity of food.

2. Identify and list the foods that provide no vitamin C and determine their source (animal or plant). What do you observe about the source of vitamin-containing foods?

3. Compare and record the number of milligrams of vitamin C in the intake of Jo Cool with the RDA. Is it too high, too low, or about right?

4. As a group delete one of this day's high-vitamin C foods and add another that would not significantly change the amount of vitamin C but would increase the amount of fiber. Use a food composition table to verify that your choice is a good one.

Name the food you are deleting. ____

- Indicate the amount of vitamin C in deleted food. _______
- Indicate the amount of fiber in the deleted food. _______

Name and give the amount of food you are adding. ______________________

- Indicate the amount of Vitamin C in added food. _______
- Indicate the amount of fiber in the added food. _______

5. List three major functions of vitamin C in the body.

6. Identify the foods eaten by Jo Cool that contain 10 or more RAEs of vitamin A. List them in order of decreasing amounts of RAE of vitamin A they contain. Also record quantity of food. If the food contains preformed vitamin A place (**PA**) after the food, if it contains beta carotene, place (**BC**) after the food, and if it has been fortified with vitamin A, place (**F**) after the food.

7. Compare and record the number of RAEs of vitamin A in this day's intake with the RDA for vitamin A. Is it too high, too low, or about right?

8. What is a vitamin A-rich fruit that would be a good replacement for the pudding dessert?

9. What is a vitamin A-rich vegetable that would be a good replacement for the corn?

10. List three major functions of vitamin A in the body.

11. For each of the following B vitamins, identify the two major sources in this day's diet.

thiamin (vit. B_1):

riboflavin (vit. B_2):

niacin (vit. B_3):

vitamin B_6:

vitamin B_{12}:

folate:

12. As a group make a general statement about the major function of the B vitamins.

13. Why do you think the corn flakes are high in many of the vitamins?

by Lorrie Miller Kohler

Name____________________________

DIET ANALYSIS ACTIVITY DA-10S: MINERAL EXERCISE

Objectives:

- To identify food sources of selected minerals.
- To use guidelines and recommendations to evaluate intake of selected minerals.
- To suggest menu choices that would provide adequate intake of selected minerals.
- To identify major functions of selected minerals.

Instructions:

Answer the following questions about minerals using the simulated computer printout for Jo Cool (pp. 111-117).

1. Review the foods eaten by Jo Cool using the Intake Spreadsheet and identify the foods that contain 10 milligrams (mg) or more of calcium. List these foods in decreasing order of amount of calcium they contain. Also record quantity of food.

Foods containing 10 mg or more of calcium	

2. Compare and record the number of milligrams of calcium in this day's intake with the DRI (AI) for Jo Cool. Is it too high, too low, or about right?

3. As a group add a food of <u>plant</u> origin that is a good source of calcium. Use a food composition table to verify that your choice is a good one.

Name and give the amount of the food you are adding. ________________________

How much calcium is there in the food you are adding? __________

4. List two major functions of calcium in the body.

5. Identify the foods eaten by Jo Cool that contain 0.7 milligrams (mg) or more of iron. List them in order of decreasing amounts of milligrams of iron they contain. Also record quantity of food.

Foods containing 0.7 mg or more of iron		

6. Compare and record the number of milligrams of iron in this day's intake with the RDA for Jo Cool. Is it too high, too low, or about right?

7. As a group change this day's intake to <u>increase</u> the amount of iron. List 3 additional food sources of iron and the amount of iron that each contains.

   1.

   2.

   3.

8. Describe the major function of iron in the body.

9. Why do the hamburger bun, doughnut and cornflakes have relatively high iron content (as compared with other foods in this day's intake)?

10. What is the total amount of sodium consumed in this day?

11. Do you think that this intake is high, low, or about right? Upon what guideline or recommendation (other than the computer printout) are you basing your opinion?

12. Carefully review the foods and their sodium content. As a group suggest three specific menu changes that would lower the sodium content of this day's intake. List the food and the amount of sodium it contains.

    1.

    2.

    3.

Group__________________

COOPERATIVE GROUP'S EVALUATION FORM

At the completion of your Base Group's activity, complete this evaluation form. The Reader/ Recorder reads the questions to the group, and records the group's agreed upon answers. Place the Evaluation Form into the group's folder and turn it in with the completed group activity.

1. Overall, how effectively did your group work together on this assignment? (Circle the appropriate response.)

1	2	3	4	5
not at all	poorly	adequately	well	extremely well

2. How many of the total number of group members participated actively most of the time? (Circle the appropriate number.)

   0 1 2 3 4

3. How many of the group's members were fully prepared for the group work most of the time? (Circle the appropriate number.)

   0 1 2 3 4

4. Each person give one specific example of something you learned from the group that you probably wouldn't have learned on your own.

5. Suggest one specific practical change the group could make that would help improve everyone's learning.

6. Describe how well each group member carried out his or her respective role.

FORMAL COOPERATIVE LEARNING GROUP ACTIVITIES

This semester you will have the opportunity to work in Formal Cooperative Learning Groups. In these groups you will work with a group of people in what is called a base group, have a specific role in the group, and follow a particular group process. The ultimate goal of formal cooperative group learning and process is to ensure the learning of each member of the group.

Early on in the semester each student will be assigned to a cooperative learning base group with three other students with whom they will work for the remainder of the semester. The task of the base group is to discuss the group activities 6I-10I. Each student's 1-day *Diet Analysis Plus* printout is to be used as the basis for answering questions found in the activities on carbohydrates, lipids, proteins, vitamins and minerals. Your instructor will provide due dates for when the activities are to be completed in class. On those dates, the questions in each activity will be discussed in the assigned base groups. Since the intent of cooperative group learning is to foster the group's learning, it is important that all members take seriously their responsibility to complete the activities. After the group discussion is finished, each group will complete a short written evaluation on how their group functioned. Then, as a class, we will summarize the answers to the questions that comprise the activity. Since the intent of the group process is that all members of the group learn the content of a particular activity, all should be prepared to answer any question if asked to do so. All group members will turn in the particular group activity and their group's evaluation form in the assigned group's folder after the discussion is completed.

Guidelines for Cooperative Groups:
The following guidelines are important to follow when working in cooperative learning groups:

1. Learn the names of your group members.
2. When your group meets, arrange chairs so you are facing each other; everyone should be able to hear and see all members and feel included.
3. Each person should have a role in the group. Determine roles before you begin the discussion of the activity. Rotate roles at each group meeting.
4. Each student is responsible to him/herself and to his/her team.
5. Each student has an obligation to learn the material and to try to ensure that all teammates learn it.

Social Skills for Cooperative Groups:
The following social skills are important to effectively facilitate the cooperative group process:

1. Tolerant Listening
2. Constructive Disagreement
3. Asking for Clarification
4. Expressing Need for Assistance
5. Summarizing

Roles and Tasks:

The following roles and tasks will be used in our cooperative groups:

1. Reader: reads aloud the questions in the exercise to the group members.
2. Checker: asks each person to state their answer to the questions and asks if all agree/disagree. All group members should agree on answer to question and understand answer.
3. Summarizer: summarizes the answer to each question after discussion of the question is completed by the group.
4. Timer: watches the time, keeps group on task and makes sure group finishes in allotted time. At the completion of the exercise, timer sees that all papers are placed in the group's folder and that ABSENTEES are recorded on the attendance sheet inside the folder. Folders are turned in to instructor.

by Lorrie Miller Kohler

Name______________________

DIET ANALYSIS ACTIVITY DA-6I: CARBOHYDRATE AND FIBER EXERCISE

Objectives:
- To identify foods high in simple and complex carbohydrates and in fiber.
- To use guidelines and recommendations to evaluate a menu (or diet).
- To suggest menu changes that would improve carbohydrate and fiber in the diet.
- To calculate percent calories from carbohydrates.

Instructions:
Answer the following questions about carbohydrates, fiber and calories using your *Diet Analysis Plus* printout.

1. Using the Intake Spreadsheet, list the foods you ate in order from highest to lowest according to amount (in grams) of carbohydrate they contain. Also record the quantity of food. Do the same for fiber containing foods.

Carbohydrate-rich foods (more than 5 grams)		Fiber-rich foods (more than 1 gram)	

2. Which foods contain little or no carbohydrate? fiber?

Little/no carbohydrate (less than 1 gram)		Little/no fiber (less than 1 gram)	

3. a. Are the foods listed in #1 which are high in carbohydrate and fiber of plant or animal origin?

   b. Are the foods listed in #2 of plant or animal origin?

   c. From questions 3a and b, what do you conclude about the sources carbohydrates and fiber?

4. Which foods contain complex carbohydrates (starch)? simple carbohydrates?

Complex carbohydrates (starch)	Simple carbohydrates

5. **Each group member** suggest a change you would make to increase the complex carbohydrates (starch) in this day's intake. Record suggestions below.

6. Would you make any changes in the fiber intake? If so, describe one. If not, why not? Record group members' changes or reasons for not changing below.

7. Select a food item and calculate the percent of calories in this food that came from carbohydrate. Be sure **all** group members understand how to do these calculations. <u>Show your math</u>.

8. Calculate what percent of the total calories came from carbohydrate. Be sure **all** group members understand how to do these calculations. <u>Show your math</u>.

9. In a sentence, describe how the percent of total calories from carbohydrate in this day's intake compares with the DRI goals.

10. Do you think this day's intake implemented the *Dietary Guidelines for Americans* related to calories, carbohydrates and fiber? (Review *Dietary Guidelines* if necessary.) If so, how? If not, how not? Record group members' responses below.

by Lorrie Miller Kohler

Name______________________

DIET ANALYSIS ACTIVITY DA-7I: LIPID EXERCISE

Objectives:
- To identify foods high in saturated and unsaturated fats, and in cholesterol.
- To calculate percent fat by weight and percent calories from fat.
- To use guidelines and recommendations to evaluate a menu for fat content.
- To suggest dietary changes that would improve fat and cholesterol intake.

Instructions:
Answer the following questions about fats, cholesterol, and calories using your 1-day *Diet Analysis Plus* printout.

1. Review the Intake Spreadsheet from the *Diet Analysis Plus* printout. Identify and list foods <u>you</u> believe high in total fat and in cholesterol.

Foods high in total fat	Foods high in cholesterol

2. a. Using the Intake Spreadsheet, identify those foods containing 3 or more grams of total fat, saturated fat, monounsaturated fat, and polyunsaturated fat, and list them in order from high to low. Also record quantity of food.

Total fat	Saturated fat	Monounsat. fat	Polyunsat. fat

b. Do the same for those foods containing 5 or more milligrams of cholesterol. Record milligrams of cholesterol and quantity of food.

Cholesterol			

3. Select a high-fat food. Calculate % fat by weight in this food by dividing the weight of the fat (in grams) by the total weight of the food (in grams) and multiply by 100.

_____ grams/_____ grams =_____ x 100 =______% fat by weight in ____________________ (selected food)

4. Calculate % calories from fat for this food.
   Step 1. Calculate the number of calories in the grams of fat this food contains.
   Step 2. Divide that number by total calories in the food.
   Step 3. Multiply by 100.

Calories from fat/total food cal. =_____ x 100% =_____ % cal. from fat in ____________________ (food)

5. Write a single sentence describing how the answer from Question #3 is different from the answer from Question #4.

6. What percent of total caloric intake for this day was contributed by this high-fat food?
   Calculate this by:
   Step 1. Multiplying grams of fat in food x 9 = calories from fat in the food.
   Step 2. Dividing this number by total number of calories for the day and multiplying by 100.

Calories from fat/total calories for day x 100 =_____% calories from fat in ____________________ (food)

7. Note how Questions 3, 4, and 6 are asking different questions. As a group agree on and summarize the differences in a single sentence and write it below.

8. What percent of the total calories come from saturated, monounsaturated, and polyunsaturated fat? The total grams of these fats is given in the Intake Spreadsheet. To calculate the amounts, multiply the number of grams of each type of fat by 9; then divide each answer by the total calories; and multiply by 100. Show your math.

   ______% calories from saturated fat

   ______% calories from monounsaturated fat

   ______% calories from polyunsaturated fat

9. How do the values in Question 8 compare with recommendations for total fat intake of 20%-35% of calories with less than 10% of calories coming from saturated fat? As a group summarize the comparisons in a sentence and write it here.

10. Would you recommend any changes in the foods containing fats that were selected in this day's intake? If yes, what suggestion(s)? If not, why not? Record group's responses below.

11. In Question #2 you listed foods containing cholesterol from in this day's intake. What is the origin—plant or animal—of these foods?

12. Select 4 foods containing cholesterol and identify the specific source of the cholesterol in each.

Cholesterol-Containing Food	Source of Cholesterol

13. How does the total cholesterol consumed in this day's intake compare with the recommendation to consume less than 300 milligrams of cholesterol a day?

14. Would you make any suggestions for change in your cholesterol intake? If yes, what suggestion(s)? If no, why not? Record changes or reasons for no change below.

by Lorrie Miller Kohler

Name______________________

DIET ANALYSIS ACTIVITY DA-8I: PROTEIN EXERCISE

Objectives:
- To identify foods high in quantity of protein.
- To evaluate the adequacy of protein intake in grams and in the context of caloric intake.
- To compare the quality of protein in foods from plant and animal sources.
- To calculate percent calories from protein.
- To suggest menu choices that are high in protein and fiber and low in fat.

Instructions:
Answer the following questions about proteins using your 1-day *Diet Analysis Plus* printout.

1. Review the foods you ate using the Intake Spreadsheet and identify the foods that contain more than 3 grams of protein. List these foods in decreasing order of amount of protein and according to their source. Also record quantity of food.

Animal-derived protein		**Plant-derived protein**	

2. Compare the quantity of protein obtained from these two sources. As a group agree on and describe in a single sentence the comparison. Record it below.

3. Describe and compare the total number of grams of protein (_____) with your RDA for protein (_____). * Remember the number of grams of protein consumed should be no more than twice the RDA for protein. What is twice the RDA for you?_____ Make this comparison in the context of looking at your calorie intake (______) for the day and your RDA for calories (______). In looking at the gram intake in the context of the caloric intake, is the quantity of your protein intake for this day adequate? Why or why not?

4. Each group **member** compare the quality of protein from the two sources (in Question #1). As a group describe the comparison in a single sentence. Record it below.

5. In the space below rank all the protein-rich foods listed in Question #1 from highest to lowest according to the amount (in grams) of total fat. Record number of grams of fat.

6. After looking at the data in Question #5, as a group, make a statement comparing the source of high protein-rich foods with their fat content.

7. What percent of the total calories for the day came from protein? Show your calculations below.

8. Do you think this day's intake implemented the DRI goals for protein? If yes, how? If no, how not? Record group members' responses.

9. As a group delete one of this day's high-protein foods and add another protein source which **would not change the amount of protein, but would decrease the amount of total fat and increase the amount of fiber.** Use a food composition table to verify that your choice is a good one.

Name the food you are deleting. ______________________

- How much protein does it contain? ________
- How much fat does it contain? ________
- How much fiber does it contain? ________

Name and give the amount of the food you are adding. ______________________

- Indicate the amount of protein in the food you are adding. ________
- Indicate the amount of fat in the food you are adding. ________
- Indicate the amount of fiber in the food you are adding. ________

by Lorrie Miller Kohler

Name____________________________

DIET ANALYSIS ACTIVITY DA-9I: VITAMIN EXERCISE

Objectives:

- To identify food sources of selected vitamins.
- To use RDAs to evaluate intake of selected vitamins.
- To suggest menu choices that would provide adequate intake of selected vitamins.
- To identify major functions of selected vitamins.

Instructions:

Answer the following questions about vitamins using your 1-day *Diet Analysis Plus* printout.

1. Review the foods you ate using the Intake Spreadsheet and identify the foods that contain 10 milligrams or more of vitamin C. List these foods in decreasing order of amount in milligrams of vitamin C they contain. Also record quantity of food.

2. Identify and list the foods that provide no vitamin C and determine their source (animal or plant). What do you observe about the source of vitamin-containing foods?

3. Compare and record the number of milligrams of vitamin C in your intake with the RDA. Is it too high, too low, or about right?

4. As a group delete one of this day's high vitamin C foods and add another that would not significantly change the amount of vitamin C but would increase the amount of fiber. Use a food composition table to verify that your choice is a good one.

Name the food you are deleting. ________________________________

- Indicate the amount of vitamin C in deleted food. ______
- Indicate the amount of fiber in the deleted food. ______

Name and give the amount of food you are adding. ________________________________

- Indicate the amount of Vitamin C in added food. ______
- Indicate the amount of fiber in the added food. ______

5. List three major functions of vitamin C in the body.

6. Identify the foods you ate that contain 10 or more RAEs of vitamin A. List them in order of decreasing amounts of RAE of vitamin A they contain. Also record quantity of food. If the food contains preformed vitamin A place (**PA**) after the food, if it contains beta carotene, place (**BC**) after the food, and if it has been fortified with vitamin A, place (**F**) after the food.

7. Compare and record the number of RAEs of vitamin A in this day's intake with the RDA for vitamin A. Is it too high, too low, or about right?

8. Select a sweet food low in vitamin A, and identify a vitamin A-rich fruit that would be a good replacement.

9. Select a low-vitamin A vegetable and identify a vitamin A-rich vegetable that would be a good replacement.

10. List three major functions of vitamin A in the body.

11. For each of the following B vitamins, identify the two major sources in this day's diet.

thiamin (vit. B_1):

riboflavin (vit. B_2):

niacin (vit. B_3):

vitamin B_6:

vitamin B_{12}:

folate:

12. As a group make a general statement about the major function of the B vitamins.

13. Did you eat any processed foods high in vitamins? Why do you think such foods are high in many of the vitamins?

by Lorrie Miller Kohler

Name______________________________

DIET ANALYSIS ACTIVITY DA-10I: MINERAL EXERCISE

Objectives:

- To identify food sources of selected minerals.
- To use guidelines and recommendations to evaluate intake of selected minerals.
- To suggest menu choices that would provide adequate intake of selected minerals.
- To identify major functions of selected minerals.

Instructions:
Answer the following questions about minerals using your 1-day *Diet Analysis Plus* printout.

1. Review the foods you ate using the Intake Spreadsheet and identify the foods that contain 10 milligrams (mg) or more of calcium. List these foods in decreasing order of amount of calcium they contain. Also record quantity of food.

Foods containing 10 mg or more of calcium	

2. Compare and record the number of milligrams of calcium in this day's intake with your DRI (AI). Is it too high, too low, or about right?

3. As a group add a food of plant origin that is a good source of calcium. Use a food composition table to verify that your choice is a good one.

Name and give the amount of the food you are adding. ______________________________

How much calcium is there in the food you are adding? __________

4. List two major functions of calcium in the body.

5. Identify the foods you ate that contain 0.7 milligrams (mg) or more of iron. List them in order of decreasing amounts of milligrams of iron they contain. Also record quantity of food.

Foods containing 0.7 mg or more of iron		

6. Compare and record the number of milligrams of iron in this day's intake with your RDA. Is it too high, too low, or about right?

7. As a group change this day's intake to increase the amount of iron. List 3 additional food sources of iron and the amount of iron that each contains.

   1.

   2.

   3.

8. Describe the major function of iron in the body.

9. Are there any foods containing refined grains? Why do such foods have relatively high iron content?

10. What is the total amount of sodium consumed in this day?

11. Do you think that this intake is high, low, or about right? Upon what guideline or recommendation (other than the computer printout) are you basing your opinion?

12. Carefully review the foods and their sodium content. As a group suggest three specific menu changes that would lower the sodium content of this day's intake. List the food and the amount of sodium it contains.

    1.

    2.

    3.

Group____________________

COOPERATIVE GROUP'S EVALUATION FORM

At the completion of your Base Group's activity, complete this evaluation form. The Reader/ Recorder reads the questions to the group, and records the group's agreed upon answers. Place the Evaluation Form into the group's folder and turn it in with the completed group activity.

1. Overall, how effectively did your group work together on this assignment? (Circle the appropriate response.)

1	2	3	4	5
not at all	poorly	adequately	well	extremely well

2. How many of the total number of group members participated actively most of the time? (Circle the appropriate number.)

   0 1 2 3 4

3. How many of the group's members were fully prepared for the group work most of the time? (Circle the appropriate number.)

   0 1 2 3 4

4. Each person give one specific example of something you learned from the group that you probably wouldn't have learned on your own.

5. Suggest one specific practical change the group could make that would help improve everyone's learning.

6. Describe how well each group member carried out his or her respective role.

Exercise #1 – Carbohydrates: Evaluating a Menu

Use the menu below to answer the following questions.

Menu for Cassy Carbohydrate	
4 pancakes (4" across, ¼ inch thick) 2 tsp. margarine (soft) 6 tsp. syrup 1 cup orange juice 1 cup skim milk Soft shell taco (large): 1 flour tortillas (10 inches across) 2 oz. lean ground beef 2 oz. mozzarella cheese 1 cup shredded romaine lettuce 1 sliced tomato (≈ ½ cup) 1 tbsp. chopped onion 3 tbsp. Taco Sauce	1 cups reduced-fat (2%) milk 5 small Vanilla Wafers 3 oz. roasted chicken breast (w/o skin) 1 large baked potato (weight 6 oz.) 2 tbsp. sour cream (fat-free) 1 cup peas (peas are "starchy" vegetables) 1 large bran muffin (weight 4 oz.) 1 cup apple juice

1. Identify the main sources of carbohydrate in this menu.

2. Divide your carbohydrate sources into foods containing mostly complex carbohydrates or mostly simple carbohydrates.

Mostly Complex Carbohydrates	Mostly Simple Carbohydrates

3. In your list of simple carbohydrates, which ones would you classify as naturally occurring sugars and which ones would you classify as refined or added sugars? Why?

4. In your list of complex carbohydrates, which ones would you classify as significant sources of dietary fiber? Why?

5. Next estimate the total grams of carbohydrate in this menu. To do this you need to determine: (a) the exchange list and (b) corresponding number of exchanges for each item.

6. Referring to the food sources of carbohydrate in this menu, what are two strengths of this menu?

7. What suggestions do you have to improve the sources of carbohydrates the in this menu? Be specific. Example: use whole fruit instead of fruit juice (for more dietary fiber).

EXERCISE #2 – CARBOHYDRATE TICK TACK TOE

Directions: Students can play in pairs or groups as preferred by the instructor. Before marking an "X" or "O" over one of the squares below, the student/group must complete the questions within that square. Students should refer to the following:

- Cassy's menu from "Carbohydrates" worksheet, p. 157
- "Carbohydrates" questions, pp. 157-158
- *Diet Analysis* + printouts for Cassy Carbohydrate, pp. 188-195

• Using the menu on p. 157, answer questions 1, 2, 3 and 4 on pp. 157-158 for the breakfast. • Define *nutrient density*	• Using the menu on p. 157, answer questions 1, 2, 3 and 4 on pp. 157-158 for the lunch and afternoon snack (cookies and milk). • Define *glycemic index*	• Using the menu on p. 157, answer questions 1, 2, 3 and 4 on pp. 157-158 for the dinner and evening snack (muffin and apple juice). • Define *available carbohydrate*
• Calculate the % of kcal from carbohydrate in this menu (p. 189). • Compare the % to the AMDR (Acceptable Macronutrient Distribution Range)	• List functions of carbohydrates. • Answer questions 6 and 7 on p. 158.	• What foods provide the most fiber in this menu? • Check your answer, using either p. 191 or p. 195. Notice serving sizes. Which format is easier to use?
• Using the list of foods on p. 157, circle foods with added sugars. Compare added sugars to total carbohydrates using a ratio (assume the carbohydrates in the circled foods are added sugars). Is this menu high or low in **added sugars**?	• Using the information on either p. 189 or p. 190-191, calculate the g of fiber/1000 kcal in this menu. • Compare this amount to the recommended intake of fiber (11.5 g/1000 kcal).	• One starch exchange provides 15 g of carbohydrate. Using the spreadsheet on p. 190, four pancakes = ≈ _____ starch exchanges; one apple bran muffin = ≈ _____ starch exchanges. • Explain these calculations.

EXERCISE #3 – DIETARY FATS (TOTAL, SAT, MONO, AND POLY) AND CHOLESTEROL

Use the menu below to answer the following questions.

Menu for Sally Sand	
2 oz. Kix cereal (≈ 1 ½ cups) 1 medium banana 1 cup reduced-fat (2%) milk 1 cup coffee 2 tsp. half and half 2 slices whole wheat toast 1 tsp. margarine (soft) Bacon/Lettuce/Tomato Sandwich: 4 slices bacon (20 slices/lb.) 1 tbsp. mayonnaise 2 leaves Romaine lettuce 1 sliced tomato (≈ ½ cup) 2 slices toasted whole wheat bread Salad: 1.5 cup Romaine lettuce (chopped) 1 oz. grated cheddar cheese 1 tbsp. grated carrots ¼ raw cucumber 1 tbsp sunflower seeds ¼ of a medium avocado ½ chopped egg 3 tbsp. Ranch dressing (reduced fat) 3 cups ice tea with lemon	16 wheat thins 2 tbsp. Creamy peanut butter Stir-fry: 3 ounce beef sirloin steak 1 cup cooked brown rice 1/3 cup broccoli 1/3 cup carrots ¼ chopped green pepper ¼ cup chopped mushrooms 1 tablespoon soy sauce 1 cup reduced-fat (2%) milk 1 cup vanilla ice cream 1 cup fresh whole strawberries, unsweetened

1. Identify the main sources of fat in this menu.

2. Classify your fat sources as sources of either invisible or visible fat. Foods classified as visible fats generally have 90% or more of their kcal from fat. Many sources of invisible fat contain 50 -80% of their kcal from fat.

Invisible Fat	Visible Fat

3. Classify your fat sources as containing predominately saturated fatty acids, monounsaturated fatty acids or polyunsaturated fatty acids.

Mostly Saturated	Mostly Monounsaturated	Mostly Polyunsaturated

4. What fat sources in this menu are high in the essential fatty acids, linoleic and linolenic? Note these are also referred to as omega-6 and omega-3 fatty acids. What other foods in this menu supply omega-3 fatty acids?

5. Identify the food sources of cholesterol in this menu.

6. Are there any sources of *trans* fatty acids in this menu? If yes, list them.

7. Using the Exchange Lists, estimate the total grams of fat in this menu.

8. In terms of the diet planning principles and dietary guidelines, what are some of the strengths of this menu?

9. Based on the detailed information about the sources and types of lipids in this menu, what specific changes would you suggest? Example: reduce number of servings of visible fats and oils (such as salad dressing).

EXERCISE #4 – USING *DIET ANALYSIS PLUS* PRINTOUTS TO EVALUATE FAT AND CHOLESTEROL INTAKE

Use the printouts from *Diet Analysis Plus* (pp. 196-203) to evaluate about Sally Sand's food record.

1. Using the Intake vs. Goals report for this menu on (p. 197), for which nutrients did Sally consume less than her recommended goals? What nutrients did Sally consume in amounts proportionate to her kcal intake of 120% of her goal? Based on this information, would you say that Sally's diet is nutrient dense? Explain.

2. Use the printout to calculate the percentage of Sally's kcal that are coming from fat. Compare to the AMDR for total fat. What is your conclusion?

3. Sally's intake of total fat, saturated fat, monounsaturated fat, polyunsaturated fat and cholesterol are presented in detail in the Intake Spreadsheet for this menu (pp. 199-200). Use Sally's goals for fat intake (found on her profile, p. 196) and divide by her intakes from the spreadsheet. Compare to the goal % found on her Intake vs. Goals report for this menu (p. 197).

4. List the five highest contributors of total fat in Sally's menu. Include the grams of total fat consumed in your answer. List servings sizes consumed by Sally. Use the printout that sorts total fat intake from most to least (p. 201). Compare the serving sizes consumed by Sally to "recommended/reference serving sizes."

5. On the printouts (total fat, polyunsaturated fat, and cholesterol) that sort intake from most to least, notice the percentages listed next to the grams (or milligrams for cholesterol). Explain what these percentages represent. Consider these percentages as a pie chart.

6. List the five highest contributors of saturated fat in Sally's menu. Include the size of portion consumed. How could you use this information to help Sally reduce her intake of saturated fat? Use the spreadsheet (pp. 199-200) for this question.

7. List the highest contributors of cholesterol in her diet. What percent of the Daily Value for cholesterol did Sally obtain? The Daily Value for cholesterol is 300 milligrams. How would you use this information? (Use p. 203.)

8. Evaluate Sally's intake of monounsaturated and polyunsaturated fatty acids as percentages of kcal. Describe your conclusions.

9. What are the main sources of polyunsaturated fatty acids in Sally's diet? (Use p. 202.)

10. What recommendations do you have for Sally? Back up your answer with specific suggestions for Sally to improve her lipid intake.

EXERCISE #5 – PROTEIN AND AMINO ACIDS REVIEW ACTIVITY

1. Using the Exchange System, estimate the amount of protein in the following meal. First, identify the exchange list and the number of "exchanges" per item. Then, determine the grams of protein per item.

   2 oz. roasted chicken breast
   ¼ cup red bell pepper
   ½ cup zucchini squash
   ½ cup soybean sprouts
   1 cup cooked brown rice
   1 ½ cups raw spinach
   1 oz. almonds
   2 tablespoons Italian dressing
   1 cup low-fat vanilla yogurt
   1/3 cup raspberries, sweetened
   1 shortbread cookie

   Total grams of protein in this meal = ______________

2. Refer to the meal above and answer the following questions.

   A. Identify the main sources of complete protein in the meal.

   B. What plant foods in this meal are significant sources of protein (2 grams or more per serving)?

   C. Lysine is a limiting essential amino acid. What plant foods in this meal are low in lysine?

   D. Identify the plant proteins in this meal that provide lysine.

   E. Methionine and tryptophan are limiting essential amino acids. What plant foods in this meal are low in these two amino acids?

   F. Identify the plant proteins in this meal that provide methionine and tryptophan.

   G. Identify the complementary plant proteins in this menu.

H. Are any of the animal proteins in this meal low in lysine, methionine or tryptophan?

3. Using your DRI recommended intake for protein, determine the percentage of your recommended intake for protein provided by this meal. Show your calculations and label your units.

4. What is the ratio of grams of animal protein to grams of total protein in this meal? Comment on the ratio (state whether it is high, low or about right).

5. What are your conclusions about this meal? What changes would you make to this meal and why?

6. Calculate the percentage of kcal from protein in this meal using the grams of protein supplied and the kcal in the meal (see printout for Paul Protein on pp. 204-206 for kcal). Compare this to the AMDR.

7. If you consume 2400 calories and 15% of your kcal come from protein, how many grams of protein is this?

   Likewise, if you consume 800 calories and 15% of your kcal come from protein, how many grams of protein is this?

   Use your answers to these two calculations to explain why it is important to know your RDA for protein as well as your goal AMDR.

EXERCISE #6 – VITAMIN CALCULATIONS

Using the *Diet Analysis Plus* printouts for Viola Vitamin (pp. 207-211), calculate the following.

1. Nutrient density ratio (% nutrient DRI goal ÷ % kcal DRI goal supplied) for vitamin A.

2. Nutrient density ratio for vitamin C.

3. In your own words, state the meaning of one of the two nutrient density ratios.

4. Are there any vitamins with nutrient density ratios less than 1? If yes, name them. If no, write none.

5. What vitamin has the highest nutrient density ratio? the lowest nutrient density ratio?

6. True or false: This diet is nutrient dense. Explain your answer.

To be considered a significant source of a nutrient a food must supply at least 25% of the RDA/AI; an excellent source must supply at least 75%; a good source supplies 50%; and a fair source supplies 10%. Use this classification system to evaluate intake of vitamins.

7. Viola's RDA for folate is 400 μg DFE (Dietary Folate Equivalents). Are there any significant sources (at least 25% of the RDA) of folate in her diet? If yes, list them. Note: the percentages on the Source Analysis report represent the % of the total for the specific nutrient supplied by a food, not the % of the RDA. These percentages are ranked from highest to lowest.

8. Use the Intake Spreadsheet (pp. 209-210) to identify the three highest sources of vitamin E in Viola's diet. Her RDA for vitamin E is 15 milligrams α- tocopherol. Use this information to classify the three sources (excellent, good, significant or fair).

9. Viola's RDA for vitamin C is 75 milligrams. Are there any excellent sources (at least 75% of the RDA) of vitamin C in her diet? If yes, list them.

10. If you take a supplement with 1 gram of vitamin C, is this considered a pharmacological dose? Calculate how many times the RDA is consumed and compare to the UL for vitamin C. Approximately how many oranges would this be?

EXERCISE #7 – MINERAL CALCULATIONS

Use the *Diet Analysis Plus* printouts for Minnie Mineral (pp. 212-218) to answer the following questions and complete the calculations. Show your work.

1. What mineral has the highest nutrient density ratio (% nutrient DRI goal ÷ % kcal DRI goal supplied)?

2. What mineral has the lowest nutrient density ratio?

3. Calculate the nutrient density ratio for magnesium.

4. Interpret the nutrient density ratio that you calculated in question 3.

5. Sodium does not have an RDA but it does have a Daily Value. Show how the % goal for sodium in the meal was calculated. Sodium now has an AI; recalculate a % goal for Minnie's intake using the AI for a female, age 29.

6. When evaluating an individual's intake of sodium, a ratio known as milligrams of sodium per kcal is often used. Calculate this ratio using the information on the printout.

7. Interpret the ratio from question 6. What is the recommended ratio?

8. This intake supplies 19.09 milligrams of iron with 1370 kcal. Express this value in milligrams per 1000 kcal. The usual intake of iron is 5-6 milligrams per 1000 kcal. Using your calculation, how does this diet compare?

The following questions use different printouts - see page reference with each question.

9. Zinc is found primarily in animal flesh and vegetarians may have lower intakes. What foods supplied zinc in Minnie's diet? Compare the amount to her RDA of 8 mg/day. Zinc is found on the Intake Spreadsheet (p. 214).

To be considered a significant source of a nutrient a food must supply at least 25% of the RDA/AI; an excellent source must supply at least 75%; a good source supplies 50%; and a fair source supplies 10%.

10. Using the RDA standard for magnesium for 19-30 year old females as a "yardstick," are there any significant sources of magnesium in this diet? Name them. Show your calculation.

11. Use the information on amount of sodium in foods in this person's intake (p. 215) and make specific suggestions to this person. When you use *Diet Analysis Plus*, you can print information on single nutrients; the Source Analysis report will rank order food sources from highest to lowest.

12. Salmon (with bones) is often listed as a significant source of calcium. This person ate a pasta salad made with canned salmon. Using the Adequate Intake guideline for calcium for young adults 19-30, is the salmon in this meal a significant source? Consider % supplied for the serving size. Show your calculation.

13. Using the Source Analysis for calcium printout (p. 216), how many calcium equivalents did Minnie's mozzarella cheese and skim milk provide?

14. Most foods are considered fair sources of iron (supplying approximately 10%) of the RDA. What are the four highest sources of iron in this diet? Evaluate these food sources in terms of your personal RDA for iron. Show your calculations.

Consider the dinner meal (picnic of pasta-salmon salad containing salmon, macaroni, mayonnaise, egg, celery and pickle relish; carrot sticks, cantaloupe, whole grain bagel, ice-cold skim milk, and brownie). Using these 11 foods, complete questions 15-19.

15. Determine the total iron in this meal and the amount of vitamin C in this dinner of 11 food items.

16. Calculate the amount of heme iron and non-heme iron in the meal. Show your calculations.

Definitions

Heme iron: 40% of the iron in foods derived from animal flesh
Nonheme iron: the remaining 60% of iron in foods derived from animal flesh; all other sources of iron including plant foods, such as grains, legumes, vegetables, nuts, seeds, and fruits; non-flesh animal foods such as milk, cheese, yogurt, and eggs; and iron salts used in enriched and fortified foods or dietary supplements.

17. About 25% of the heme iron will be absorbed and 17% of the nonheme iron; therefore, multiply your heme iron from above by 0.25 and your nonheme iron by 0.17. Add these two values to get total iron absorbed. Show your work and circle your answer.

    What do you conclude about this meal?

18. What would be the effect of not including vitamin C and the MFP factor in a meal (as far as amount of iron absorbed)? Prepare a list of suggestions to individuals to increase the amount of iron absorbed in their diet.

19. Compare mineral needs for Minnie Mineral when she is pregnant to her non-pregnant needs. Use Minnie's Profile DRI Goals reports (pp. 217-218).

EXERCISE #8A – RECIPE EVALUATION AND MODIFICATION USING TABLES OF FOOD COMPOSITION, DIETARY REFERENCE INTAKES, AND AMDR

Learning Objectives – After completing this project you will be able to:

- Analyze a recipe
- Use tables of food composition
- Calculate % of kcal from carbohydrate, fat and protein (for a recipe)
- Determine personal nutrient standards for kcal and protein
- Look up and use Daily Values for cholesterol, sodium and fiber
- Look up and use DRI for vitamins and minerals
- Suggest ways to "modify" a recipe using nutrient standards
- Re-analyze the same recipe using your suggested modifications in ingredients
- Write-up your "results"

Steps and Calculations:

1. Select recipe (this will be done in class). Your instructor may provide an assortment of traditional main dish and entrée salad recipes to choose from, or direct you to bring a recipe to class for the instructor's approval.

2. Using a table of food composition and the Nutrient Spreadsheet Form (p. 177-178), analyze the recipe (this process continues from steps 2 through 4). To do this you look up the nutrient information in the table of food composition and enter on the form (adjusting amounts as described in step 3 of these directions).

   If you cannot find an ingredient in the table, go to the following USDA website: http://www.nal.usda.gov/fnic/foodcomp/search/ or check with your instructor for additional resources.

   Food label data for this assignment may be insufficient (reason = information on vitamins and minerals is often incomplete and values are expressed as % instead of mg and micrograms). However, if you need to use food labels, ask your instructor to assist you in using the label information to obtain accurate values for nutrients supplied.

   Hint to make your life easier: Type the information into a spreadsheet program; you can set it up to perform the calculations for you.

   Tips to using food composition data:
   - Keep the same number of decimal places as found in the food composition table.
   - Do not round values off.
   - Values listed as < 1, < .1, and < .01 indicate negligible amounts; record as "trace."
   - Obtain complete nutrient information for all ingredients; do not leave any blanks in your spreadsheet.
   - If you do not plan to type and print out the information from the Nutrient Spreadsheet Form using a word processor or spreadsheet program, write legibly.

3. **Adjust data from the food composition table as needed** when analyzing the recipe (increasing or decreasing amounts used). For example, if the nutrient data in the table is for 3 oz. cooked hamburger

and the recipe calls for 1 lb. raw hamburger, determine approximate oz. cooked (in this case 12 oz.) and multiply the data in the table by 4.

- If a recipe specifies to serve with rice or pasta, include enough for the entire recipe.
- If a recipe lists optional ingredients, it is your choice whether to use or not to include them in your analysis.

4. After you have entered the nutrient data, **total all columns on your Nutrient Spreadsheet** and **divide by the number of servings to obtain information per serving**. Double-check the math (this is where a spreadsheet helps out).

- The rest of this project will use the nutrient information per serving.

---

5. Use the carbohydrate, fat and protein grams supplied by one serving from your spreadsheet to calculate the % of kcal per serving from carbohydrate, fat and protein. Include your calculations in the space below and write your answers in the box.

Hint: multiply grams by kcal value/gram; add up the total kcal; use this kcal total to calculate your percentages. Otherwise your % will not add up to 100%.
Example: % of kcal from fat = $\dfrac{(\text{g fat} \times 9)}{(\text{g carb.} \times 4) + (\text{g fat} \times 9) + (\text{g protein} \times 4)}$

Percentage of kcal from carbohydrate, fat and protein per serving – Original Recipe		
% kcal from carbohydrate	% kcal from fat	% kcal from protein

Suggestion: Show this information in a pie chart (generated by the spreadsheet program) and attach.

In your write up (step 10) discuss these percentages. This information is useful for making decisions on how to modify the recipe.

6. Calculate or look up the following values. Place your answers in the space below (in order) and label your answers.

   a. Estimate your energy (kcal) requirements using the appropriate formula or table from your textbook. (Check with your instructor if you are not sure which method to use.)

b. Calculate your recommended protein intake based on the recommendation of 0.8 g/kg of body weight/day for healthy adults: **weight (kg) × 0.8** or **[weight (lbs.) ÷ 2.2] × 0.8.**

c. Calculate your recommended intake ranges for carbohydrate and fat using the AMDR for these nutrients and your estimated energy needs (from 6a above). Multiply the minimum and maximum percentages (expressed as decimals; e.g. 50% = 0.50) by the kcal needs to determine the desired kcal for each. Then, divide each number by the kcal per gram for that nutrient to determine your recommended intake range in grams.

d. Calculate your personalized recommendation for fiber based on your estimated total energy needs (from 6a above). The FDA set the Daily Value for fiber at 25 grams or 11.5 g per 1000 kcal, while the DRI recommendation for fiber is slightly higher at 14 g per 1000 kcal. You may use either the FDA or DRI recommendation for this calculation.

e. Look up the Daily Value for cholesterol in your textbook.

f. The recommendation for saturated fat is that it contributes **no more than** 10% of total kcal. Calculate your recommended intake of saturated fat: less than or equal to 10% (0.10) times your estimated kcal needs (from 6a above), divided by 9 kcal/g.

g. For your remaining nutrients, use the DRI tables for vitamins and minerals from your text. Notice the units (mg and μg) in column headings. Vitamins and minerals are in separate tables.

7. Enter your recommended values for 5a-5g on the Dietary Analysis Form (p. 179) in the "standard" row, under the appropriate column headings.

8. Enter the amounts in one serving of your recipe on the "intake or value" row of the Dietary Analysis Form and calculate % supplied ("intake as a % of standard"). The formula is the standard % supplied formula or: **(amount in recipe ÷ standard) × 100%** (units must match). These percentages will allow you to evaluate the overall nutrient density of the recipe and identify strengths and weaknesses (vitamins, minerals, and protein).

9. The last part of this assignment involves evaluating your results for steps 2-8 and modifying your recipe to meet selected personal goals (such as more protein, less fat, more calcium or iron, etc.). Since these are main dish recipes, modifying ingredients (using more vegetables or low-fat cheese; changing ingredients; using less added fat; substituting more nutrient dense ingredients) isn't generally a problem.

Make a several modifications (based on your personal goals and preferences) and re-analyze your recipe using a second Nutrient Spreadsheet and Dietary Analysis Form (pp. 180-182), following the same steps 2-5 and 7-8 with your modified recipe as you completed for your original recipe.

After calculating the % kcal from carbohydrate, fat and protein in your modified recipe as explained in step 5 above, show your calculations in the space below and enter your new percentages in the table below:

Percentage of kcal from carbohydrate, fat and protein per serving – Modified Recipe		
% kcal from carbohydrate	% kcal from fat	% kcal from protein

Suggestion: Show this information in a pie chart (generated by the spreadsheet program) and attach.

Compare these percentages to the original recipe and explain your modifications in your written summary (step 10).

<u>Writing Up Your Results</u>:

10. Collect all your information (this assignment with calculations completed on pages 174-176, the original and modified recipe, Nutrient Spreadsheets, and Dietary Analysis Forms). You may wish to make a table of contents and/or include tabs for various parts of the assignment.

    Prepare a two- to three-page typed summary of this assignment. Include specific examples in your summary from your spreadsheets, forms, calculations and/or tables. Indicate other applications of this project than you envision using or encountering (personally or professionally). Attach this summary to the rest of your information and hand in!

    Remember to include your recipe with your project. Indicate recipe modifications either directly on your recipe or on a separate sheet of paper. Don't forget to clearly explain the reasoning you used in the modifications and the outcome (analyzed results) in your summary.

Nutrient Spreadsheet Form (page 1 of 2) – Original Recipe

Food	Weight/ measure	kcal	Prot. (g)	Carb. (g)	Fiber (g)	Fat (g)	Sat. fat (g)	Mono. fat (g)	Poly. fat (g)	Chol (mg)	Calc. (mg)	Iron (mg)
Total												

Nutrient Spreadsheet Form (page 2 of 2) – Original Recipe

Food	Magn. (mg)	Pota. (mg)	Sodi. (mg)	Zinc (mg)	Vit. A RAE (µg)	Thia. (mg)	Vit. E (mg)	Ribo. (mg)	Niac. NE (mg)	Vit. B_6 (mg)	Fol. DFE (µg)	Vit. C (mg)
Total												

Dietary Analysis Form – Original Recipe

	kcal	Prot. (g)	Carb. (g)	Fiber (g)	Fat (g)	Sat. fat (g)	Mono. fat (g)	Poly. fat (g)	Chol (mg)	Calc. (mg)	Iron (mg)
Intake or value											
Standard							---	---			
Intake as a % of standard											

	Magn. (mg)	Pota. (mg)	Sodi. (mg)	Zinc (mg)	Vit. A RAE (µg)	Thia. (mg)	Vit. E (mg)	Ribo. (mg)	Niac. NE (mg)	Vit. B_6 (mg)	Fol. DFE (µg)	Vit. C (mg)
Intake or value												
Standard												
Intake as a % of standard												

Nutrient Spreadsheet Form (page 1 of 2) – Modified Recipe

Food	Weight/ measure	kcal	Prot. (g)	Carb. (g)	Fiber (g)	Fat (g)	Sat. fat (g)	Mono. fat (g)	Poly. fat (g)	Chol (mg)	Calc. (mg)	Iron (mg)
Total												

Nutrient Spreadsheet Form (page 2 of 2) – Modified Recipe

Food	Magn. (mg)	Pota. (mg)	Sodi. (mg)	Zinc (mg)	Vit. A RAE (µg)	Thia. (mg)	Vit. E (mg)	Ribo. (mg)	Niac. NE (mg)	Vit. B_6 (mg)	Fol. DFE (µg)	Vit. C (mg)
Total												

Dietary Analysis Form – Modified Recipe

	kcal	Prot. (g)	Carb. (g)	Fiber (g)	Fat (g)	Sat. fat (g)	Mono. fat (g)	Poly. fat (g)	Chol (mg)	Calc. (mg)	Iron (mg)
Intake or value											
Standard							---	---			
Intake as a % of standard											

	Magn. (mg)	Pota. (mg)	Sodi. (mg)	Zinc (mg)	Vit. A RAE (µg)	Thia. (mg)	Vit. E (mg)	Ribo. (mg)	Niac. NE (mg)	Vit. B_6 (mg)	Fol. DFE (µg)	Vit. C (mg)
Intake or value												
Standard												
Intake as a % of standard												

by Elaine M. Long

EXERCISE #8B – RECIPE EVALUATION AND MODIFICATION USING *DIET ANALYSIS PLUS*

Learning Objectives – After completing this project you will be able to:

- Analyze a recipe
- Use the Create a Recipe feature in *Diet Analysis Plus*
- Determine % of kcal from carbohydrate, fat and protein (for a recipe)
- Determine personal nutrient standards for kcal and protein using Profile DRI Goals
- Suggest ways to "modify" a recipe using nutrient standards
- Re-analyze the same recipe using your suggested modifications in ingredients
- Write-up your "results"

Steps and Reports:

1. Select recipe (this will be done in class). Your instructor may provide an assortment of traditional main dish and entrée salad recipes to choose from, or direct you to bring a recipe to class for the instructor's approval.

2. Set up your personal profile in *Diet Analysis Plus,* if you haven't already done so. Your DRI goals will be used as the standards for comparing the nutrient content of the original and modified recipes.

3. Using *Diet Analysis Plus,* enter the recipe.

   a. Click the Track Diet button; then, beside "Select a day," click on a date for which you haven't entered any intake information (you can scroll to previous or future weeks if necessary).

   b. Click the Create a Recipe button. In the Create a New Recipe box, enter the recipe name and the total number of portions the recipe makes. Click the Create a Recipe button to begin adding ingredients.

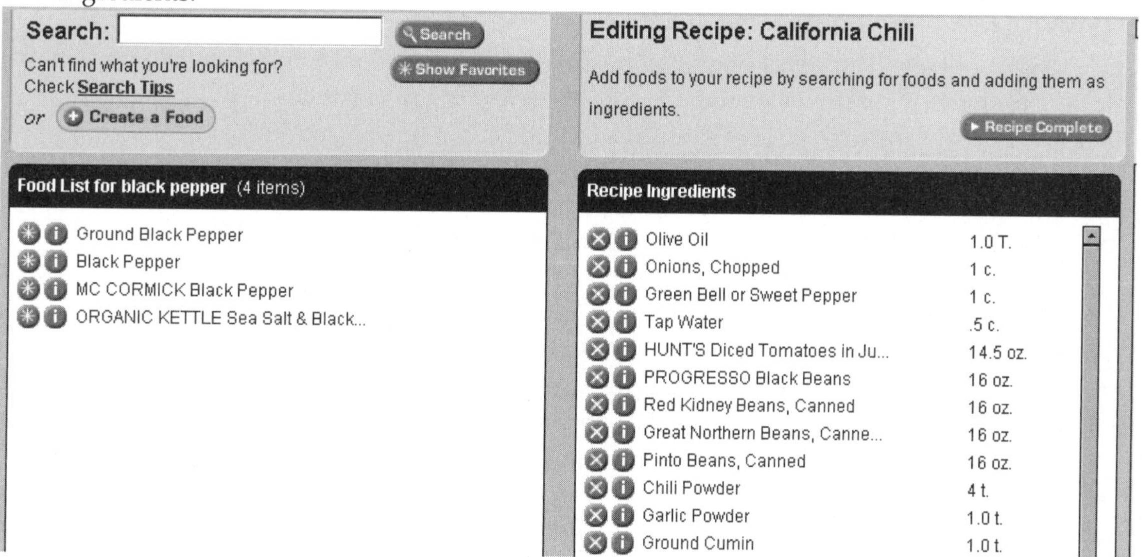

   c. Search for each ingredient, and specify the appropriate amount to add to the recipe. This is done in the same manner as adding a day's food/beverage intake. (See chili example above.)

d. Click the Recipe Complete button. You will return to the Track Diet screen.

Tips for entering the recipe:

- Take care to enter the correct amount and units for each ingredient when building the recipe.
- If a recipe specifies to serve with rice or pasta, include enough for the entire recipe.
- If a recipe lists optional ingredients, it is your choice whether to use or not to include them in your analysis.

4. On the Track Diet screen, click the Show Favorites button. The recipe you just entered will be on your favorites list. Click the name of the recipe to add it to the daily intake for your chosen date, and select 1 piece as the amount—this will allow you to analyze 1 serving of the recipe.

   Click the Create Reports button to display the list of reports you can view and print (on the right under "Report Type"). You can click on the names of the reports to view the analysis of 1 portion of your recipe, making sure you choose the same date beside "From date" and "To date."

   - The rest of this project will use the nutrient information per serving.

---

5. View and print the Macronutrient Ranges report. Fill in the table below using the percentages of kcal from each energy nutrient from this report.

Percentage of kcal from carbohydrate, fat and protein per serving – Original Recipe		
% kcal from carbohydrate	% kcal from fat	% kcal from protein

In your write up (step 9) discuss these percentages. This information is useful for making decisions on how to modify the recipe.

6. View and print your Profile DRI Goals report using *Diet Analysis Plus*. Enter your DRI for each nutrient from this report in the "standard" row of the Dietary Analysis Form (p. 186).

   The recommendation for saturated fat is that it contributes **no more than** 10% of total kcal. Calculate your recommended intake of saturated fat: less than or equal to 10% (0.10) times your kcal needs (from the report), divided by 9 kcal/g. Enter this as the "standard" for saturated fat.

7. View and print the Intake vs. Goals report, making sure to choose the same date where you entered your recipe as a food eaten as the "From date" and "To date." Using this report, enter the amounts of each nutrient in one serving of your recipe on the "intake or value" row of the Dietary Analysis Form. Fill in the "intake as a % of standard" row using the goal percentages.

   Refer to the Macronutrient Ranges report and use the percentage of kcal from carbohydrate and fat as the "intake as a % of standard" for these nutrients. You will need to calculate the percentage for saturated fat, by dividing the grams of saturated fat (from the Intake vs. Goals report) by the standard you calculated previously.

These percentages will allow you to evaluate the overall nutrient density of the recipe and identify strengths and weaknesses (vitamins, minerals, and protein).

8. The last part of this assignment involves evaluating your results for steps 2-7 and modifying your recipe to meet selected personal goals (such as more protein, less fat, more calcium or iron, etc.). Since these are main dish recipes, modifying ingredients (using more vegetables or low-fat cheese; changing ingredients; using less added fat; substituting more nutrient dense ingredients) isn't generally a problem.

   Make a several modifications (based on your personal goals and preferences) and re-analyze your recipe using a second Dietary Analysis Form (p. 187). Begin by clicking the Track Diet button and choosing a different date than you used for the original recipe; then follow steps 3b-7 with your modified recipe.

   Enter the percentage of kcal from each energy nutrient for the modified recipe (as determined in step 5) in the table below.

Percentage of kcal from carbohydrate, fat and protein per serving – Modified Recipe		
% kcal from carbohydrate	% kcal from fat	% kcal from protein

   Compare these percentages to the original recipe and explain your modifications in your written summary (step 9).

<u>**Writing Up Your Results:**</u>

9. Collect all your information (this assignment with tables completed on pages 184-185, the original and modified recipe, *Diet Analysis Plus* printouts, and Dietary Analysis Forms). You may wish to make a table of contents and/or include tabs for various parts of the assignment.

   Prepare a two- to three-page typed summary of this assignment. Include specific examples in your summary from your spreadsheets, forms, calculations and/or tables. Indicate other applications of this project than you envision using or encountering (personally or professionally). Attach this summary to the rest of your information and hand in!

   Remember to include your recipe with your project. Indicate recipe modifications either directly on your recipe or on a separate sheet of paper. Don't forget to clearly explain the reasoning you used in the modifications and the outcome (analyzed results) in your summary.

Dietary Analysis Form – Original Recipe

	kcal	Prot. (g)	Carb. (g)	Fiber (g)	Fat (g)	Sat. fat (g)	Mono. fat (g)	Poly. fat (g)	Chol (mg)	Calc. (mg)	Iron (mg)
Intake or value											
Standard							---	---			
Intake as a % of standard											

	Magn. (mg)	Pota. (mg)	Sodi. (mg)	Zinc (mg)	Vit. A RAE (µg)	Thia. (mg)	Vit. E (mg)	Ribo. (mg)	Niac. NE (mg)	Vit. B_6 (mg)	Fol. DFE (µg)	Vit. C (mg)
Intake or value												
Standard												
Intake as a % of standard												

Dietary Analysis Form – Modified Recipe

	kcal	Prot. (g)	Carb. (g)	Fiber (g)	Fat (g)	Sat. fat (g)	Mono. fat (g)	Poly. fat (g)	Chol (mg)	Calc. (mg)	Iron (mg)
Intake or value											
Standard							---	---			
Intake as a % of standard											

	Magn. (mg)	Pota. (mg)	Sodi. (mg)	Zinc (mg)	Vit. A RAE (μg)	Thia. (mg)	Vit. E (mg)	Ribo. (mg)	Niac. NE (mg)	Vit. B_6 (mg)	Fol. DFE (μg)	Vit. C (mg)
Intake or value												
Standard												
Intake as a % of standard												

Profile

Profile Name	Cassy Carbohydrate
Height	5 ft. 6 inches
Weight	134 lb.
Age	25 years
BMI	22

DRI Goals

Nutrient	DRI	
Energy Nutrients		
Calories (kcal)	2182 kcal	
Carbohydrates	245 - 355 g	*45%-65% of kilocalories*
Fat	48 - 85 g	*20%-35% of kilocalories*
Protein	55 - 191 g	*10%-35% of kilocalories*
Protein (g/kg)	49 g	*Daily requirement based on grams per kilogram of body weight*
Fat		
Saturated Fat	24	*less than 10% of calories recommended*
Monounsaturated Fat	-	*No recommendation*
Polyunsaturated Fat	-	*No recommendation*
Cholesterol	300 mg	*less than 300mg recommended*
Essential fatty acids		
PFA 18:2, Linoleic	12 g	
PFA 18:3, Linolenic	1.1 g	
carbohydrates		
Dietary Fiber, Total	25 g	
Sugar, Total	-	*No recommendation*
Other		
Water	2.3 L	
Alcohol	-	*No recommendation*
Vitamins		
Thiamin (Vit B1)	1.1 mg	
Riboflavin	1.1 mg	
Niacin	14 mg	
Vitamin B6	1.3 IU	
Vitamin B12	2.4 µg	
Folate (DFE)	400 µg	
Vitamin C	75 mg	
Vitamin D (ug)	5 µg	
Vitamin A(RAE)	699.9 µg	
alpha-tocopherol (Vit E)	15 mg	
Minerals		
Calcium	1000 mg	*DRI Adequate Intake*
Iron	18 mg	
Magnesium	310 mg	
Potassium	4700 mg	*DRI Adequate Intake*
Zinc	8 mg	
Sodium	1500 mg	*DRI Adequate Intake*

! Nutrient	Intake	DRI	0% 50% 100%
Energy			
Calories	2477 kCal	2181.5 kCal	114%
Carbohydrates	353 g	245 g - 355 g	
Fat	68 g	48 g - 85 g	
Protein	113 g	55 g - 191 g	
Protein(g/kg/day)	113 g	49 g	231%
Fat			
Saturated Fat	25.57 g	*no rec*	
Monounsaturated Fat	24.42 g *	*no rec*	
Polyunsaturated Fat	10.14 g *	*no rec*	
Cholesterol	303 mg	300 mg	101%
Essential fatty acids			
PFA 18:2, Linoleic	6.4 g *	12 g	53%
PFA 18:3, Linolenic	0.94 g *	1.1 g	85%
carbohydrates			
Dietary Fiber, Total	25.4 g	25 g	102%
Sugar, Total	159.01 g *	*no rec*	
Other			
Water	1.53 L *	2.3 L	67%
Alcohol	0 g *	*no rec*	
Vitamins			
Thiamin (Vit B1)	1.88 mg *	1.1 mg	171%
Riboflavin	2.33 mg *	1.1 mg	212%
Niacin	25.99 mg *	14 mg	186%
Vitamin B6	2.28 IU *	1.3 IU	175%
Vitamin B12	6.48 µg *	2.4 µg	270%
Folate (DFE)	252.26 µg *	400 µg	63%
Vitamin C	172.75 mg	75 mg	230%
Vitamin D (ug)	4.94 µg *	5 µg	99%
Vitamin A(RAE)	3043.2 µg	699.9 µg	435%
alpha-tocopherol (Vit E)	3.81 mg *	15.0 mg	25%
Minerals			
Calcium	1521.59 mg	1000 mg	152%
Iron	15.44 mg	18 mg	86%
Magnesium	283.95 mg *	310 mg	92%
Potassium	3758.04 mg *	4700 mg	80%
Zinc	12.74 mg *	8 mg	159%
! Sodium	2707.29 mg	1500 mg	180%

* This nutrient may be under-represented as not all foods in the selected report range have a value for this nutrient.

Item Name	*Quantity*	*Weight*	Kcal (kcal)	Protein (g)
Mozzarella Cheese...	2.0 oz.	0 g	171.23	14.71
Red Tomato, Diced	0.50 c.	90 g	16.20	0.79
Vanilla Wafer Coo...	6.0 item	36 g	170.27	1.54
Potatoes, Flesh a...	1.0 item	202 g	220.17	4.64
Pancakes, Prepare...	4.0 item	152 g	331.36	11.85
Green Peas, Froze...	1.0 c.	160 g	124.80	8.23
Pancake Syrup	6.0 t.	38.40 g	89.85	0
Apple Juice, Unsw...	1.0 c.	248 g	116.55	0.14
Nonfat, Skim or F...	1.0 c.	245 g	83.30	8.25
Ground Beef, Regu...	2.0 oz.	0 g	165.46	15.41
Romaine Lettuce, ...	1.0 c.	56 g	9.51	0.68
Onions, Chopped	1.0 t.	10 g	4.19	0.9
Reduced Fat Milk,...	1.0 c.	244 g	122	8.5
Chicken Breast, M...	3.0 oz.	0 g	140.25	26.36
Sour Cream	2.0 t.	24 g	51.36	0.75
Orange Juice, Pre...	1.0 c.	259.38 g	116.71	1.76
ORTEGA Medium Tac...	3.0 t.	48 g	30	0
PARKAY Margarine,...	2.0 t.	8.96 g	64	0
McDONALD'S Low Fa...	1.0 item	114 g	300	6
LA TORTILLA Low F...	1.0 item	51 g	150	4
Totals:			**2477.21 kcal**	**114.51 g**

Item Name	Carb (g)	Fat (g)	Sat Fat (g)	Mono Fat (g)
Mozzarella Cheese...	2.17	11.35	7.18	3.24
Red Tomato, Diced	3.52	0.18	0.4	0.4
Vanilla Wafer Coo...	25.59	6.98	1.77	3.98
Potatoes, Flesh a...	50.96	0.20	0.5	0
Pancakes, Prepare...	43.92	11.70	3.10	3.14
Green Peas, Froze...	22.81	0.43	0.7	0.3
Pancake Syrup	23.60	0	0	0
Apple Juice, Unsw...	28.96	0.27	0.4	0.1
Nonfat, Skim or F...	12.15	0.19	0.28	0.11
Ground Beef, Regu...	0	11.2	4.33	4.82
Romaine Lettuce, ...	1.84	0.16	0.2	0
Onions, Chopped	1.1	0	0	0
Reduced Fat Milk,...	11.41	4.80	2.34	2.4
Chicken Breast, M...	0	3.3	0.85	1.5
Sour Cream	1.2	5.3	3.13	1.45
Orange Juice, Pre...	27.96	0.15	0.1	0.2
ORTEGA Medium Tac...	6	0	0	0
PARKAY Margarine,...	0	7.4	1.28	3.41
McDONALD'S Low Fa...	61	3	0.50	-
LA TORTILLA Low F...	29	2.50	0.50	1
Totals:	**353.19 g**	**69.11 g**	**27.56 g**	**26.5 g**

Item Name	Poly Fat (g)	Chol (mg)	Linoleic (g)	Linolenic (g)
Mozzarella Cheese...	0.35	30.61	0.25	0.10
Red Tomato, Diced	0.12	0	0.11	0
Vanilla Wafer Coo...	0.87	0	0.83	0.4
Potatoes, Flesh a...	0.8	0	0.6	0.2
Pancakes, Prepare...	4.44	107.92	3.91	0.48
Green Peas, Froze...	0.20	0	0.16	0.3
Pancake Syrup	0	0	0	0
Apple Juice, Unsw...	0.8	0	0.6	0.1
Nonfat, Skim or F...	0.1	4.90	0	0

Ground Beef, Regu...	0.41	57.23	0.30	0.4
Romaine Lettuce, ...	0.8	0	0.2	0.6
Onions, Chopped	0	0	0	0
Reduced Fat Milk,...	0.16	19.52	0.1	0.1
Chicken Breast, M...	0.65	72.25	0.50	0.2
Sour Cream	0.18	10.56	0.11	0.7
Orange Juice, Pre...	0.3	0	0.2	0
ORTEGA Medium Tac...	0	0	0	0
PARKAY Margarine,...	1.40	0	-	-
McDONALD'S Low Fa...	-	0	-	-
LA TORTILLA Low F...	1	0	-	-
Totals:	**12.58 g**	**302.99 mg**	**7.87 g**	**3.58 g**

Item Name	Diet Fiber (g)	Sugar (g)	Water (L)	Alcohol (g)
Mozzarella Cheese...	0	0.34	0.2	0
Red Tomato, Diced	1.8	2.36	0.8	0
Vanilla Wafer Coo...	0.72	-	0	0
Potatoes, Flesh a...	4.24	3.23	0.14	0
Pancakes, Prepare...	2.82	8.34	0.8	0
Green Peas, Froze...	8.80	7.44	0.12	0
Pancake Syrup	0.26	23.31	0.1	0
Apple Juice, Unsw...	0.24	27.3	0.21	0
Nonfat, Skim or F...	0	12.47	0.22	0
Ground Beef, Regu...	0	0	0.2	0
Romaine Lettuce, ...	1.17	0.66	0.5	0
Onions, Chopped	0.14	0.42	0	0
Reduced Fat Milk,...	0	12.34	0.21	0
Chicken Breast, M...	0	0	0.5	0
Sour Cream	0	0.3	0.1	0
Orange Juice, Pre...	0.90	27.49	0.22	0
ORTEGA Medium Tac...	0	1.50	-	-
PARKAY Margarine,...	0	0	0	0
McDONALD'S Low Fa...	3	32	-	0
LA TORTILLA Low F...	2	0	-	0
Totals:	**26.9 g**	**159.50 g**	**4.32 L**	**0 g**

Item Name	Vit B1 (mg)	Ribo (mg)	Niacin (mg)	Vit B6 (IU)
Mozzarella Cheese...	0.5	0.18	0.6	0.4
Red Tomato, Diced	0.3	0.1	0.53	0.7
Vanilla Wafer Coo...	0.12	0.7	1.7	0
Potatoes, Flesh a...	0.21	0.6	3.32	0.70
Pancakes, Prepare...	0.30	0.47	1.87	0.15
Green Peas, Froze...	0.45	0.15	2.36	0.18
Pancake Syrup	0	0	0	0
Apple Juice, Unsw...	0.5	0.4	0.24	0.7
Nonfat, Skim or F...	0.11	0.44	0.23	0.9
Ground Beef, Regu...	0.2	0.11	3.66	0.16
Romaine Lettuce, ...	0.4	0.3	0.17	0.4
Onions, Chopped	0	0	0	0.1
Reduced Fat Milk,...	0.9	0.45	0.22	0.9
Chicken Breast, M...	0.5	0.9	11.65	0.50
Sour Cream	0	0.3	0.1	0
Orange Juice, Pre...	0.20	0.4	0.52	0.11
ORTEGA Medium Tac...	-	-	-	-
PARKAY Margarine,...	0	0	0	0
McDONALD'S Low Fa...	-	-	-	-

LA TORTILLA Low F...	0.9	0.6	0	-
Totals:	**5.59 mg**	**6.10 mg**	**27.17 mg**	**5.90 IU**

Item Name	Vit B12 (µg)	Fol (DFE) (µg)	Vit C (mg)	Vit D (ug) (µg)
Mozzarella Cheese...	1.30	5.67	0	-
Red Tomato, Diced	0	13.50	11.43	0
Vanilla Wafer Coo...	0.1	24.47	0	-
Potatoes, Flesh a...	0	-	26.5	0
Pancakes, Prepare...	0.51	-	0.91	-
Green Peas, Froze...	0	94.40	15.84	0
Pancake Syrup	0	0	0	0
Apple Juice, Unsw...	0	0	2.23	0
Nonfat, Skim or F...	1.29	12.25	0	2.45
Ground Beef, Regu...	1.85	5.66	0	0
Romaine Lettuce, ...	0	76.15	13.43	0
Onions, Chopped	0	1.89	0.63	0
Reduced Fat Milk,...	1.12	12.19	0.48	2.44
Chicken Breast, M...	0.28	3.40	0	0
Sour Cream	0.7	2.64	0.21	0.5
Orange Juice, Pre...	0	-	100.89	0
ORTEGA Medium Tac...	-	-	0	-
PARKAY Margarine,...	0	-	0	-
McDONALD'S Low Fa...	-	-	0.60	-
LA TORTILLA Low F...	0	-	0	0
Totals:	**7.15 µg**	**252.22 µg**	**173.15 mg**	**5.39 µg**

Item Name	Vit A (RAE) (µg)	alpha-T (mg)	Calcium (mg)	Iron (mg)
Mozzarella Cheese...	88.82	0.20	414.47	0.14
Red Tomato, Diced	227.15	0.27	9	0.24
Vanilla Wafer Coo...	0.10	-	9	0.79
Potatoes, Flesh a...	0	0.8	20.20	2.74
Pancakes, Prepare...	115.14	1.37	326.80	1.97
Green Peas, Froze...	1018.8	0.21	38.40	2.43
Pancake Syrup	0	0	1.15	0.1
Apple Juice, Unsw...	0.75	0.1	17.36	0.91
Nonfat, Skim or F...	151.43	0.7	222.94	1.22
Ground Beef, Regu...	0	0.10	6.79	1.55
Romaine Lettuce, ...	985.33	0.19	18.47	0.54
Onions, Chopped	0.6	0.2	2.20	0.1
Reduced Fat Milk,...	139.73	0.13	270.83	0.24
Chicken Breast, M...	5.15	0.18	12.75	0.88
Sour Cream	47.12	0.10	27.84	0.1
Orange Juice, Pre...	61.30	0.18	23.34	0.25
ORTEGA Medium Tac...	90.90	-	0	0
PARKAY Margarine,...	96.96	0.62	0	0
McDONALD'S Low Fa...	15.15	-	100	1.44
LA TORTILLA Low F...	0	-	0	0
Totals:	**3044.43 µg**	**5.35 mg**	**1521.54 mg**	**15.64 mg**

Item Name	Magn (mg)	Potas (mg)	Zinc (mg)	Sodium (mg)
Mozzarella Cheese...	14.74	53.86	1.77	299.37
Red Tomato, Diced	9.89	213.30	0.15	4.50
Vanilla Wafer Coo...	4.31	38.51	0.11	110.16
Potatoes, Flesh a...	54.54	844.35	0.64	16.15

Pancakes, Prepare...	33.43	302.48	1.14	767.60
Green Peas, Froze...	35.20	176	1.7	115.19
Pancake Syrup	0.76	5.76	0.3	31.48
Apple Juice, Unsw...	7.44	295.11	0.7	7.44
Nonfat, Skim or F...	22.4	237.64	2.8	107.80
Ground Beef, Regu...	12.46	185.30	3.29	52.70
Romaine Lettuce, ...	7.84	138.31	0.12	4.47
Onions, Chopped	1	14.39	0.1	0.30
Reduced Fat Milk,...	26.84	341.60	1.17	114.68
Chicken Breast, M...	24.64	217.60	0.85	62.90
Sour Cream	2.64	34.56	0.6	12.72
Orange Juice, Pre...	25.93	492.81	0.12	2.59
ORTEGA Medium Tac...	-	-	-	375
PARKAY Margarine,...	0.17	6.40	0	67.20
McDONALD'S Low Fa...	-	160	-	380
LA TORTILLA Low F...	-	-	-	175
Totals:	**284.23 mg**	**3757.98 mg**	**15.56 mg**	**2707.25 mg**

Item Name	Amount	Carb		0% 25% 50% 75% 100%
McDONALD'S Low Fa...	1.0 item	61 g	★	17%
Potatoes, Flesh a...	1.0 item	51 g	★	14%
Pancakes, Prepare...	4.0 item	44 g	★	12%
LA TORTILLA Low F...	1.0 item	29 g		8%
Apple Juice, Unsw...	1.0 c.	29 g		8%
Orange Juice, Pre...	1.0 c.	28 g		8%
Vanilla Wafer Coo...	6.0 item	26 g		7%
Pancake Syrup	6.0 t.	24 g		7%
Green Peas, Froze...	1.0 c.	23 g		6%
Nonfat, Skim or F...	1.0 c.	12 g		3%
Reduced Fat Milk,...	1.0 c.	11 g		3%
ORTEGA Medium Tac...	3.0 t.	6 g		2%
Red Tomato, Diced	0.5 c.	4 g		1%
Mozzarella Cheese...	2.0 oz.	2 g		1%
Romaine Lettuce, ...	1.0 c.	2 g		1%
Sour Cream	2.0 t.	1 g		0%
Onions, Chopped	1.0 t.	1 g		0%
Ground Beef, Regu...	2.0 oz.	0 g		0%
Chicken Breast, M...	3.0 oz.	0 g		0%
PARKAY Margarine,...	2.0 t.	0 g		0%
Totals:		353 g		

Item Name	Amount	Diet Fiber		0% 25% 50% 75% 100%
Green Peas, Froze...	1.0 c.	9 g	☆	35%
Potatoes, Flesh a...	1.0 item	4 g	☆	17%
McDONALD'S Low Fa...	1.0 item	3 g	☆	12%
Pancakes, Prepare...	4.0 item	3 g	☆	11%
LA TORTILLA Low F...	1.0 item	2 g		8%
Romaine Lettuce, ...	1.0 c.	1 g		5%
Red Tomato, Diced	0.5 c.	1 g		4%
Orange Juice, Pre...	1.0 c.	0.9 g		4%
Vanilla Wafer Coo...	6.0 item	0.72 g		3%
Pancake Syrup	6.0 t.	0.26 g		1%
Apple Juice, Unsw...	1.0 c.	0.24 g		1%
Onions, Chopped	1.0 t.	0.14 g		1%
Mozzarella Cheese...	2.0 oz.	0 g		0%
Nonfat, Skim or F...	1.0 c.	0 g		0%
Ground Beef, Regu...	2.0 oz.	0 g		0%
Reduced Fat Milk,...	1.0 c.	0 g		0%
Chicken Breast, M...	3.0 oz.	0 g		0%
Sour Cream	2.0 t.	0 g		0%
ORTEGA Medium Tac...	3.0 t.	0 g		0%
PARKAY Margarine,...	2.0 t.	0 g		0%
Totals:		25 g		

Profile

Profile Name	Sally Sand
Height	5 ft. 6 inches
Weight	134 lb.
Age	25 years
BMI	22

DRI Goals

Nutrient	DRI	
Energy Nutrients		
Calories (kcal)	2182 kcal	
Carbohydrates	245 - 355 g	*45%-65% of kilocalories*
Fat	48 - 85 g	*20%-35% of kilocalories*
Protein	55 - 191 g	*10%-35% of kilocalories*
Protein (g/kg)	49 g	*Daily requirement based on grams per kilogram of body weight*
Fat		
Saturated Fat	24	*less than 10% of calories recommended*
Monounsaturated Fat	-	*No recommendation*
Polyunsaturated Fat	-	*No recommendation*
Cholesterol	300 mg	*less than 300mg recommended*
Essential fatty acids		
PFA 18:2, Linoleic	12 g	
PFA 18:3, Linolenic	1.1 g	
carbohydrates		
Dietary Fiber, Total	25 g	
Sugar, Total	-	*No recommendation*
Other		
Water	2.3 L	
Alcohol	-	*No recommendation*
Vitamins		
Thiamin (Vit B1)	1.1 mg	
Riboflavin	1.1 mg	
Niacin	14 mg	
Vitamin B6	1.3 IU	
Vitamin B12	2.4 µg	
Folate (DFE)	400 µg	
Vitamin C	75 mg	
Vitamin D (ug)	5 µg	
Vitamin A(RAE)	699.9 µg	
alpha-tocopherol (Vit E)	15 mg	
Minerals		
Calcium	1000 mg	*DRI Adequate Intake*
Iron	18 mg	
Magnesium	310 mg	
Potassium	4700 mg	*DRI Adequate Intake*
Zinc	8 mg	
Sodium	1500 mg	*DRI Adequate Intake*

Nutrient	Intake	DRI	0% 50% 100%
Energy			
Calories	2615 kCal	2181.5 kCal	120%
Carbohydrates	308 g	245 g - 355 g	
Fat	113 g	48 g - 85 g	
Protein	106 g	55 g - 191 g	
Protein(g/kg/day)	106 g	49 g	216%
Fat			
Saturated Fat	36.31 g	*no rec*	
Monounsaturated Fat	41.18 g	*no rec*	
Polyunsaturated Fat	21.95 g	*no rec*	
Cholesterol	330.75 mg	300 mg	110%
Essential fatty acids			
PFA 18:2, Linoleic	19.27 g *	12 g	161%
PFA 18:3, Linolenic	1.6 g *	1.1 g	145%
carbohydrates			
Dietary Fiber, Total	34.92 g	25 g	140%
Sugar, Total	120.76 g	*no rec*	
Other			
Water	2.38 L	2.3 L	103%
Alcohol	0.26 g *	*no rec*	
Vitamins			
Thiamin (Vit B1)	2.52 mg	1.1 mg	229%
Riboflavin	3.55 mg *	1.1 mg	323%
Niacin	30.6 mg *	14 mg	219%
Vitamin B6	3.52 IU *	1.3 IU	271%
Vitamin B12	8.92 µg *	2.4 µg	372%
! Folate (DFE)	1036.96 µg *	400 µg	259%
Vitamin C	192.77 mg	75 mg	257%
Vitamin D (ug)	8.01 µg *	5 µg	160%
Vitamin A(RAE)	4581.99 µg	699.9 µg	655%
alpha-tocopherol (Vit E)	11.47 mg *	15.0 mg	76%
Minerals			
Calcium	1525.97 mg	1000 mg	153%
Iron	29.42 mg	18 mg	163%
! Magnesium	549.66 mg *	310 mg	177%
Potassium	4309.34 mg	4700 mg	92%
Zinc	23.91 mg *	8 mg	299%
! Sodium	4407.99 mg	1500 mg	294%

* This nutrient may be under-represented as not all foods in the selected report range have a value for this nutrient.

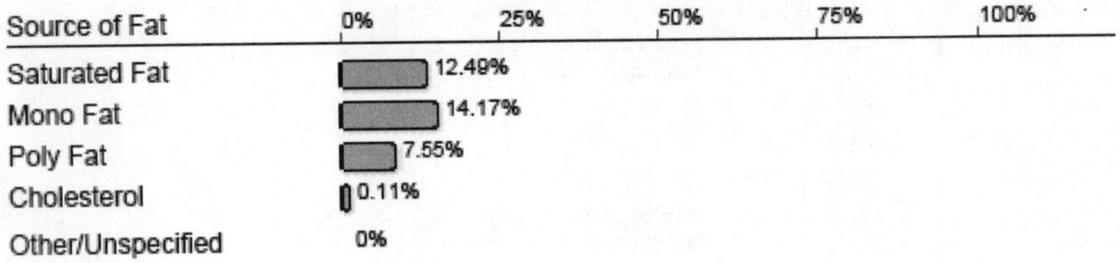

Source of Fat	0% – 100%
Saturated Fat	12.49%
Mono Fat	14.17%
Poly Fat	7.55%
Cholesterol	0.11%
Other/Unspecified	0%

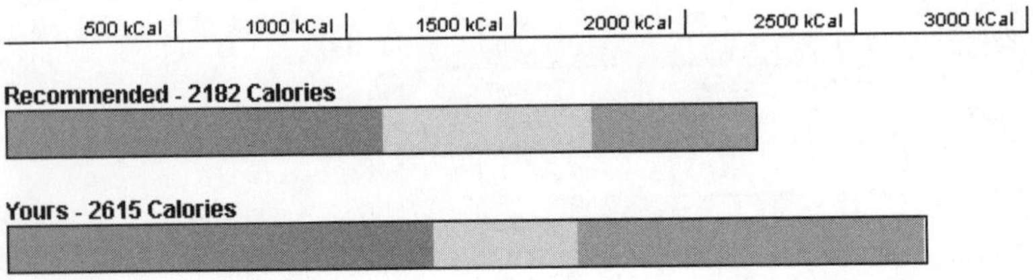

500 kCal | 1000 kCal | 1500 kCal | 2000 kCal | 2500 kCal | 3000 kCal

Recommended - 2182 Calories

Yours - 2615 Calories

	Recommended		Yours		In range?
Carbohydrates	45%-65%	982-1418 kCal	46%	1207 kCal	Yes
Fats	20%-35%	436-764 kCal	38%	991 kCal	No
Proteins	10%-35%	218-764 kCal	16%	416 kCal	Yes

- Carbohydrates
- Protein
- Fat
- Alcohol

Item Name	Quantity	Weight	Kcal (kcal)	Protein (g)
Pork Bacon, Cured...	4.0 sl.	25.20 g	136.33	9.33
Long Grain Brown ...	1.0 c.	195 g	216.44	5.3
Iceberg Lettuce L...	2.0 pc.	16 g	1.60	0.12
Carrots	0.33 c.	40.42 g	16.57	0.37
Whole Wheat Bread...	2.0 sl.	50 g	138.50	5.44
Mushrooms	0.25 c.	17.50 g	3.85	0.54
Cheddar Cheese, S...	1.0 oz.	28.35 g	114.25	7.5
Reduced Fat Milk,...	1.0 c.	244 g	122	8.5
Cucumber	0.25 item	75.25 g	11.28	0.48
Strawberries	1.0 c.	144 g	46.8	0.96
Avocado, Californ...	0.25 item	42.40 g	70.81	0.83
Reduced Fat Milk,...	1.0 c.	244 g	122	8.5
Half and Half Cre...	2.0 t.	9.60 g	12.48	0.28
Margarine, Soft	1.0 t.	4.51 g	32.30	0.3
Beef Top Sirloin,...	3.0 oz.	0 g	157.82	24.94
Mayonnaise with S...	1.0 t.	13.80 g	98.94	0.15
Broccoli	0.33 c.	29.30 g	9.96	0.82
Red Tomato, Diced	0.50 c.	90 g	16.20	0.79
Sweet Green Peppe...	0.25 item	18.50 g	4.99	0.16
Romaine Lettuce, ...	1.50 c.	84 g	14.27	1.3
Soy Sauce	1.0 t.	18 g	9.53	0.93
Carrots, Grated	1.0 t.	6.88 g	2.81	0.6
Vanilla Ice Cream	1.0 c.	132 g	265.32	4.61
Dried Sunflower S...	1.0 t.	9 g	51.29	2.5
Banana	1.0 item	118 g	105.1	1.28
Hard Boiled Egg	0.50 item	25 g	38.75	3.14
Brewed Coffee	1.0 c.	237 g	9.47	0.33
Lemon Flavored In...	3.0 c.	714 g	14.28	0
Whole Wheat Bread...	2.0 sl.	50 g	138.50	5.44
Peanut Butter, Sm...	2.0 t.	32 g	191.67	7.99
GENERAL MILLS KIX...	2.0 oz.	56.70 g	226.80	3.78
KRAFT FREE Fat Fr...	3.0 t.	52.50 g	75	0
WHEAT THINS Origi...	16.0 item	29 g	140.19	2
Totals:			**2616.10 kcal**	**109.21 g**

Item Name	Carb (g)	Fat (g)	Sat Fat (g)	Mono Fat (g)
Pork Bacon, Cured...	0.36	10.52	3.46	4.66
Long Grain Brown ...	44.77	1.75	0.35	0.63
Iceberg Lettuce L...	0.33	0.1	0	0
Carrots	3.87	0.9	0.1	0
Whole Wheat Bread...	25.85	2.40	0.51	0.94
Mushrooms	0.56	0.5	0	0
Cheddar Cheese, S...	0.36	9.39	5.97	2.66
Reduced Fat Milk,...	11.41	4.80	2.34	2.4
Cucumber	2.73	0.8	0.2	0
Strawberries	11.5	0.43	0.2	0.6
Avocado, Californ...	3.66	6.53	0.90	4.15
Reduced Fat Milk,...	11.41	4.80	2.34	2.4
Half and Half Cre...	0.41	1.10	0.68	0.31
Margarine, Soft	0.2	3.62	0.62	1.28
Beef Top Sirloin,...	0	5.64	2.19	2.40
Mayonnaise with S...	0.53	10.79	1.64	2.70
Broccoli	1.94	0.10	0.1	0
Red Tomato, Diced	3.52	0.18	0.4	0.4
Sweet Green Peppe...	1.18	0.3	0	0
Romaine Lettuce, ...	2.76	0.25	0.3	0.1
Soy Sauce	1.52	0.1	0	0

Carrots, Grated	0.65	0.1	0	0
Vanilla Ice Cream	31.15	14.52	8.96	3.92
Dried Sunflower S...	1.68	4.46	0.46	0.85
Banana	26.95	0.38	0.13	0.3
Hard Boiled Egg	0.28	2.65	0.81	1.1
Brewed Coffee	0	1.80	0	0
Lemon Flavored In...	2.85	0	0	0
Whole Wheat Bread...	25.85	2.40	0.51	0.94
Peanut Butter, Sm...	5.89	16.72	3.20	7.91
GENERAL MILLS KIX...	49.14	0.94	0	0
KRAFT FREE Fat Fr...	16.50	0	0	0
WHEAT THINS Origi...	19.6	6	0.99	2.50
Totals:	**309.41 g**	**114.97 g**	**37.36 g**	**43.15 g**

Item Name	Poly Fat (g)	Chol (mg)	Linoleic (g)	Linolenic (g)
Pork Bacon, Cured...	1.14	27.71	1	0.5
Long Grain Brown ...	0.62	0	0.60	0.2
Iceberg Lettuce L...	0	0	0	0
Carrots	0.4	0	0.4	0
Whole Wheat Bread...	0.56	0	0.53	0.2
Mushrooms	0.2	0	0.2	0
Cheddar Cheese, S...	0.26	29.76	0.16	0.10
Reduced Fat Milk,...	0.16	19.52	0.1	0.1
Cucumber	0.3	0	0	0
Strawberries	0.22	0	0.12	0.9
Avocado, Californ...	0.85	0	0.70	0.5
Reduced Fat Milk,...	0.16	19.52	0.1	0.1
Half and Half Cre...	0.4	3.55	0.2	0.1
Margarine, Soft	1.56	0	1.51	0.4
Beef Top Sirloin,...	0.21	61.34	0.17	0
Mayonnaise with S...	5.88	5.24	5.19	0.68
Broccoli	0.1	0	0	0
Red Tomato, Diced	0.12	0	0.11	0
Sweet Green Peppe...	0.1	0	0.1	0
Romaine Lettuce, ...	0.13	0	0.3	0.9
Soy Sauce	0	0	0	0
Carrots, Grated	0	0	0	0
Vanilla Ice Cream	0.59	58.8	0.36	0.22
Dried Sunflower S...	2.94	0	2.93	0
Banana	0.8	0	0.5	0.3
Hard Boiled Egg	0.35	106	0.29	0
Brewed Coffee	0	0	0	0
Lemon Flavored In...	0	0	0	0
Whole Wheat Bread...	0.56	0	0.53	0.2
Peanut Butter, Sm...	4.76	0	4.73	0.2
GENERAL MILLS KIX...	0	0	0	0
KRAFT FREE Fat Fr...	0	0	0	0
WHEAT THINS Origi...	0.49	0	-	-
Totals:	**23.86 g**	**331.44 mg**	**20.83 g**	**5.60 g**

Item Name	Diet Fiber (g)	Sugar (g)	Water (L)	Alcohol (g)
Pork Bacon, Cured...	0	0	0	0
Long Grain Brown ...	3.50	0.68	0.14	0
Iceberg Lettuce L...	0.16	0.28	0.1	0
Carrots	1.21	1.83	0.3	0
Whole Wheat Bread...	3.70	11.15	0.1	0

Item Name	Amount	Fat	0% 25% 50% 75% 100%
Peanut Butter, Sm...	2.0 t.	17 g	15%
Vanilla Ice Cream	1.0 c.	15 g	13%
Mayonnaise with S...	1.0 t.	11 g	10%
Pork Bacon, Cured...	4.0 sl.	11 g	9%
Cheddar Cheese, S...	1.0 oz.	9 g	8%
Avocado, Californ...	0.25 item	7 g	6%
WHEAT THINS Origi...	16.0 item	6 g	5%
Beef Top Sirloin,...	3.0 oz.	6 g	5%
Reduced Fat Milk,...	1.0 c.	5 g	4%
Reduced Fat Milk,...	1.0 c.	5 g	4%
Dried Sunflower S...	1.0 t.	4 g	4%
Margarine, Soft	1.0 t.	4 g	3%
Hard Boiled Egg	0.5 item	3 g	2%
Whole Wheat Bread...	2.0 sl.	2 g	2%
Whole Wheat Bread...	2.0 sl.	2 g	2%
Brewed Coffee	1.0 c.	2 g	2%
Long Grain Brown ...	1.0 c.	2 g	2%
Half and Half Cre...	2.0 t.	1 g	1%
GENERAL MILLS KIX...	2.0 oz.	0.94 g	1%
Strawberries	1.0 c.	0.43 g	0%
Banana	1.0 item	0.38 g	0%
Romaine Lettuce, ...	1.5 c.	0.25 g	0%
Red Tomato, Diced	0.5 c.	0.18 g	0%
Broccoli	0.333 c.	0.1 g	0%
Carrots	0.333 c.	0.09 g	0%
Cucumber	0.25 item	0.08 g	0%
Mushrooms	0.25 c.	0.05 g	0%
Sweet Green Peppe...	0.25 item	0.03 g	0%
Iceberg Lettuce L...	2.0 pc.	0.01 g	0%
Soy Sauce	1.0 t.	0.01 g	0%
Carrots, Grated	1.0 t.	0.01 g	0%
Lemon Flavored In...	3.0 c.	0 g	0%
KRAFT FREE Fat Fr...	3.0 t.	0 g	0%
Totals:		113 g	

Item Name	Amount	Poly Fat	0% 25% 50% 75% 100%
Mayonnaise with S...	1.0 t.	6 g	27%
Peanut Butter, Sm...	2.0 t.	5 g	22%
Dried Sunflower S...	1.0 t.	3 g	13%
Margarine, Soft	1.0 t.	2 g	7%
Pork Bacon, Cured...	4.0 sl.	1 g	5%
Avocado, Californ...	0.25 item	0.85 g	4%
Long Grain Brown ...	1.0 c.	0.62 g	3%
Vanilla Ice Cream	1.0 c.	0.59 g	3%
Whole Wheat Bread...	2.0 sl.	0.56 g	3%
Whole Wheat Bread...	2.0 sl.	0.56 g	3%
WHEAT THINS Origi...	16.0 item	0.49 g	2%
Hard Boiled Egg	0.5 item	0.35 g	2%
Cheddar Cheese, S...	1.0 oz.	0.26 g	1%
Strawberries	1.0 c.	0.22 g	1%
Beef Top Sirloin,...	3.0 oz.	0.21 g	1%
Reduced Fat Milk,...	1.0 c.	0.16 g	1%
Reduced Fat Milk,...	1.0 c.	0.16 g	1%
Romaine Lettuce, ...	1.5 c.	0.13 g	1%
Red Tomato, Diced	0.5 c.	0.12 g	1%
Banana	1.0 item	0.08 g	0%
Carrots	0.333 c.	0.04 g	0%
Half and Half Cre...	2.0 t.	0.04 g	0%
Cucumber	0.25 item	0.03 g	0%
Mushrooms	0.25 c.	0.02 g	0%
Broccoli	0.333 c.	0.01 g	0%
Sweet Green Peppe...	0.25 item	0.01 g	0%
Iceberg Lettuce L...	2.0 pc.	0 g	0%
Soy Sauce	1.0 t.	0 g	0%
Carrots, Grated	1.0 t.	0 g	0%
Brewed Coffee	1.0 c.	0 g	0%
Lemon Flavored In...	3.0 c.	0 g	0%
GENERAL MILLS KIX...	2.0 oz.	0 g	0%
KRAFT FREE Fat Fr...	3.0 t.	0 g	0%
Totals:		22 g	

Item Name	Amount	Chol	0% 25% 50% 75% 100%
Hard Boiled Egg	0.5 item	106 mg	32%
Beef Top Sirloin,...	3.0 oz.	61 mg	19%
Vanilla Ice Cream	1.0 c.	58 mg	18%
Cheddar Cheese, S...	1.0 oz.	30 mg	9%
Pork Bacon, Cured...	4.0 sl.	28 mg	8%
Reduced Fat Milk,...	1.0 c.	20 mg	6%
Reduced Fat Milk,...	1.0 c.	20 mg	6%
Mayonnaise with S...	1.0 t.	5 mg	2%
Half and Half Cre...	2.0 t.	4 mg	1%
Long Grain Brown ...	1.0 c.	0 mg	0%
Iceberg Lettuce L...	2.0 pc.	0 mg	0%
Carrots	0.333 c.	0 mg	0%
Whole Wheat Bread...	2.0 sl.	0 mg	0%
Mushrooms	0.25 c.	0 mg	0%
Cucumber	0.25 item	0 mg	0%
Strawberries	1.0 c.	0 mg	0%
Avocado, Californ...	0.25 item	0 mg	0%
Margarine, Soft	1.0 t.	0 mg	0%
Broccoli	0.333 c.	0 mg	0%
Red Tomato, Diced	0.5 c.	0 mg	0%
Sweet Green Peppe...	0.25 item	0 mg	0%
Romaine Lettuce, ...	1.5 c.	0 mg	0%
Soy Sauce	1.0 t.	0 mg	0%
Carrots, Grated	1.0 t.	0 mg	0%
Dried Sunflower S...	1.0 t.	0 mg	0%
Banana	1.0 item	0 mg	0%
Brewed Coffee	1.0 c.	0 mg	0%
Lemon Flavored In...	3.0 c.	0 mg	0%
Whole Wheat Bread...	2.0 sl.	0 mg	0%
Peanut Butter, Sm...	2.0 t.	0 mg	0%
GENERAL MILLS KIX...	2.0 oz.	0 mg	0%
KRAFT FREE Fat Fr...	3.0 t.	0 mg	0%
WHEAT THINS Origi...	16.0 item	0 mg	0%
Totals:		331 mg	

Nutrient (Paul Protein)	Intake	DRI	%
Energy			
Calories	960 kCal	2816.0 kCal	34%
Carbohydrates	121 g	317 g - 458 g	
Fat	35 g	63 g - 110 g	
Protein	49 g	70 g - 246 g	
Protein(g/kg/day)	49 g	65 g	75%
Fat			
Saturated Fat	6.18 g *	*no rec*	
Monounsaturated Fat	15.15 g *	*no rec*	
Polyunsaturated Fat	10.15 g *	*no rec*	
Cholesterol	62.01 mg	300 mg	21%
Essential fatty acids			
PFA 18:2, Linoleic	9.36 g *	17 g	55%
PFA 18:3, Linolenic	0.7 g *	1.6 g	44%
carbohydrates			
Dietary Fiber, Total	16.56 g	38 g	44%
Sugar, Total	60.39 g	*no rec*	
Other			
Water	0.61 L	3.3 L	18%
Alcohol	0 g	*no rec*	
Vitamins			
Thiamin (Vit B1)	0.58 mg	1.2 mg	48%
Riboflavin	1.14 mg	1.3 mg	88%
Niacin	13.98 mg	16 mg	87%
Vitamin B6	1.21 IU	1.3 IU	93%
Vitamin B12	1.49 µg	2.4 µg	62%
Folate (DFE)	137.74 µg *	400 µg	34%
Vitamin C	111.23 mg	90 mg	124%
Vitamin D (ug)	0 µg *	5 µg	0%
Vitamin A(RAE)	1217.35 µg	899.9 µg	135%
alpha-tocopherol (Vit E)	9.69 mg *	15.0 mg	65%
Minerals			
Calcium	616.04 mg	1000 mg	62%
Iron	7.97 mg	8 mg	100%
Magnesium	314.1 mg	400 mg	79%
Potassium	1688.57 mg	4700 mg	36%
Zinc	6.07 mg	11 mg	55%
Sodium	805.22 mg	1500 mg	54%

* This nutrient may be under-represented as not all foods in the selected report range have a value for this nutrient.

Item Name	Quantity	Weight	Kcal (kcal)	Protein (g)
Italian Salad Dre...	2.0 t.	29.40 g	85.55	0.11
Red Raspberries, ...	0.33 c.	83.25 g	85.74	0.58
Chicken Breast, M...	2.0 oz.	0 g	93.50	17.57
Zucchini Summer S...	0.50 c.	56.50 g	9.4	0.68
Long Grain Brown ...	1.0 c.	195 g	216.44	5.3
Almonds, Dry Roas...	1.0 oz.	28.35 g	169.24	6.26
Low Fat Vanilla Y...	1.0 c.	245 g	208.25	12.7
Shortbread Cookie...	1.0 item	8 g	40.15	0.48
Red Bell or Sweet...	0.25 c.	37.25 g	9.68	0.36
Soybeans, Sproute...	0.50 c.	47 g	38.6	3.98
Spinach, Trimmed ...	1.50 c.	48 g	4.73	1.34
Totals:			**961.28 kcal**	**49.36 g**

Item Name	Carb (g)	Fat (g)	Sat Fat (g)	Mono Fat (g)
Italian Salad Dre...	3.6	8.34	1.31	1.85
Red Raspberries, ...	21.77	0.13	0	0.1
Chicken Breast, M...	0	2.2	0.57	0.70
Zucchini Summer S...	1.89	0.10	0.2	0
Long Grain Brown ...	44.77	1.75	0.35	0.63
Almonds, Dry Roas...	5.46	14.97	1.14	9.54
Low Fat Vanilla Y...	33.81	3.6	1.97	0.84
Shortbread Cookie...	5.15	1.92	0.48	1.7
Red Bell or Sweet...	2.24	0.11	0.2	0
Soybeans, Sproute...	3.6	2.9	0.28	0.47
Spinach, Trimmed ...	0.7	0.14	-	-
Totals:	**122.99 g**	**36.16 g**	**6.50 g**	**15.83 g**

Item Name	Poly Fat (g)	Chol (mg)	Linoleic (g)	Linolenic (g)
Italian Salad Dre...	3.80	0	3.38	0.41
Red Raspberries, ...	0.7	0	0.4	0.2
Chicken Breast, M...	0.43	48.16	0.33	0.1
Zucchini Summer S...	0.4	0	0.1	0.2
Long Grain Brown ...	0.62	0	0.60	0.2
Almonds, Dry Roas...	3.58	0	3.58	0
Low Fat Vanilla Y...	0.8	12.25	0.6	0.2
Shortbread Cookie...	0.25	1.60	0.24	0.1
Red Bell or Sweet...	0.5	0	0.3	0.2
Soybeans, Sproute...	1.18	0	1.4	0.13
Spinach, Trimmed ...	-	0	-	-
Totals:	**12.26 g**	**62.1 mg**	**10.93 g**	**1.74 g**

Item Name	Diet Fiber (g)	Sugar (g)	Water (L)	Alcohol (g)
Italian Salad Dre...	0	2.44	0.1	0
Red Raspberries, ...	3.66	18.11	0.6	0
Chicken Breast, M...	0	0	0.3	0
Zucchini Summer S...	0.62	0.97	0.5	0
Long Grain Brown ...	3.50	0.68	0.14	0
Almonds, Dry Roas...	3.34	1.38	0	0
Low Fat Vanilla Y...	0	33.81	0.19	0
Shortbread Cookie...	0.14	1.20	0	0
Red Bell or Sweet...	0.74	1.56	0.3	0
Soybeans, Sproute...	0.37	0.20	0.3	0
Spinach, Trimmed ...	4.15	0	0.4	0

Item Name	Amount	Protein		0% 25% 50% 75% 100%
Chicken Breast, M...	2.0 oz.	18 g	★	36%
Low Fat Vanilla Y...	1.0 c.	12 g	★	25%
Almonds, Dry Roas...	1.0 oz.	6 g	★	13%
Long Grain Brown ...	1.0 c.	5 g	★	10%
Soybeans, Sproute...	0.5 c.	4 g		8%
Spinach, Trimmed ...	1.5 c.	1 g		3%
Zucchini Summer S...	0.5 c.	0.68 g		1%
Red Raspberries, ...	0.333 c.	0.58 g		1%
Shortbread Cookie...	1.0 item	0.48 g		1%
Red Bell or Sweet...	0.25 c.	0.36 g		1%
Italian Salad Dre...	2.0 t.	0.11 g		0%
Totals:		49 g		

Profile

Profile Name	Viola Vitamin
Height	5 ft. 6 inches
Weight	138 lb.
Age	31 years
BMI	22

DRI Goals

Nutrient	DRI	
Energy Nutrients		
Calories (kcal)	2159 kcal	
Carbohydrates	243 - 351 g	*45%-65% of kilocalories*
Fat	48 - 84 g	*20%-35% of kilocalories*
Protein	54 - 189 g	*10%-35% of kilocalories*
Protein (g/kg)	50 g	*Daily requirement based on grams per kilogram of body weight*
Fat		
Saturated Fat	24	*less than 10% of calories recommended*
Monounsaturated Fat	-	*No recommendation*
Polyunsaturated Fat	-	*No recommendation*
Cholesterol	300 mg	*less than 300mg recommended*
Essential fatty acids		
PFA 18:2, Linoleic	12 g	
PFA 18:3, Linolenic	1.1 g	
Carbohydrates		
Dietary Fiber, Total	21 g	
Sugar, Total	-	*No recommendation*
Other		
Water	2.7 L	
Alcohol	-	*No recommendation*
Vitamins		
Thiamin (Vit B1)	1.1 mg	
Riboflavin	1.1 mg	
Niacin	14 mg	
Vitamin B6	1.3 IU	
Vitamin B12	2.4 µg	
Folate (DFE)	400 µg	
Vitamin C	75 mg	
Vitamin D (ug)	5 µg	
Vitamin A(RAE)	699.9 µg	
alpha-tocopherol (Vit E)	15 mg	
Minerals		
Calcium	1000 mg	*DRI Adequate Intake*
Iron	18 mg	
Magnesium	320 mg	
Potassium	4700 mg	*DRI Adequate Intake*
Zinc	8 mg	
Sodium	1500 mg	*DRI Adequate Intake*

!	Nutrient	Intake	DRI	0% 50% 100%
	Energy			
	Calories	1639 kCal	2159.1 kCal	76%
	Carbohydrates	260 g	243 g - 351 g	
	Fat	47 g	48 g - 84 g	
	Protein	57 g	54 g - 189 g	
	Protein(g/kg/day)	57 g	50 g	114%
	Fat			
	Saturated Fat	12.3 g	*no rec*	
	Monounsaturated Fat	17.96 g *	*no rec*	
	Polyunsaturated Fat	8.8 g *	*no rec*	
	Cholesterol	50.22 mg	300 mg	17%
	Essential fatty acids			
	PFA 18:2, Linoleic	6.98 g *	12 g	58%
	PFA 18:3, Linolenic	0.59 g *	1.1 g	54%
	carbohydrates			
	Dietary Fiber, Total	29.41 g	21 g	140%
	Sugar, Total	100.09 g	*no rec*	
	Other			
	Water	1.15 L *	2.7 L	43%
	Alcohol	0 g	*no rec*	
	Vitamins			
	Thiamin (Vit B1)	1.93 mg *	1.1 mg	175%
	Riboflavin	2.46 mg *	1.1 mg	224%
	Niacin	22.92 mg *	14 mg	164%
	Vitamin B6	2.23 IU *	1.3 IU	172%
	Vitamin B12	4.31 µg *	2.4 µg	180%
!	Folate (DFE)	1225.4 µg *	400 µg	306%
	Vitamin C	363.51 mg	75 mg	485%
	Vitamin D (ug)	4.33 µg *	5 µg	87%
	Vitamin A(RAE)	6832.27 µg	699.9 µg	976%
	alpha-tocopherol (Vit E)	10.86 mg *	15.0 mg	72%
	Minerals			
	Calcium	1160.64 mg	1000 mg	116%
	Iron	27.62 mg	18 mg	153%
	Magnesium	327.64 mg *	320 mg	102%
	Potassium	3028.26 mg *	4700 mg	64%
	Zinc	15.79 mg *	8 mg	197%
!	Sodium	2524.85 mg	1500 mg	168%

* This nutrient may be under-represented as not all foods in the selected report range have a value for this nutrient.

KRAFT Singles Swi...	0	1.34	-	0
GENERAL MILLS KIX...	1.89	5.67	0	0
GENERAL FOODS INT...	1	0	-	0
PILLSBURY Cinnamo...	0	5	0	0
Totals:	**29.52 g**	**101.31 g**	**6.63 L**	**0 g**

Item Name	Vit B1 (mg)	Ribo (mg)	Niacin (mg)	Vit B6 (IU)
Romaine Lettuce, ...	0.6	0.5	0.26	0.6
Low Calorie Blue ...	0	0.3	0.1	0
Red Bell or Sweet...	0.2	0.3	0.36	0.10
Tofu, Firm, with ...	0.5	0.5	0	0.3
Lowfat Milk, 1%	0.4	0.45	0.22	0.9
Broccoli, Chopped...	0.4	0.9	0.43	0.15
Onions, Chopped	0.2	0.1	0.3	0.7
Whole Wheat Bread...	0.15	0.10	1.94	0.9
Mandarin Oranges,...	0.6	0.5	0.56	0.5
Red Tomato	0.1	0	0.18	0.2
Dried Coconut, Sh...	0	0	0.2	0.1
Almonds, Dry Roas...	0.2	0.24	1.9	0.3
Avocado, Californ...	0.3	0.6	0.81	0.12
Carrots	0.5	0.4	0.79	0.11
Peach	0.2	0.3	0.78	0.2
Strawberries	0.1	0.1	0.27	0.3
Long Grain White ...	0.43	0.3	2.45	0.3
Red Bell or Sweet...	0.3	0.5	0.58	0.17
White Granulated ...	0	0	0	0
Mushrooms, Boiled...	0.3	0.14	2.8	0.4
LOUIS RICH DELI T...	-	-	-	-
DEL MONTE Pineapp...	-	-	-	-
PEPPERIDGE FARM B...	0.3	0.2	0.26	-
KRAFT Singles Swi...	-	0.9	-	-
GENERAL MILLS KIX...	0.70	0.80	9.45	0.94
GENERAL FOODS INT...	-	-	-	-
PILLSBURY Cinnamo...	0.3	0.2	0.26	-
Totals:	**6.78 mg**	**8.33 mg**	**24.90 mg**	**7.29 IU**

Item Name	Vit B12 (μg)	Fol (DFE) (μg)	Vit C (mg)	Vit D (ug) (μg)
Romaine Lettuce, ...	0	114.23	20.15	0
Low Calorie Blue ...	0.7	-	0.9	-
Red Bell or Sweet...	0	6.70	70.77	0
Tofu, Firm, with ...	0	18.71	0.11	0
Lowfat Milk, 1%	1.7	12.19	0	2.44
Broccoli, Chopped...	0	84.23	50.62	0
Onions, Chopped	0	9.11	3.7	0
Whole Wheat Bread...	0	19.50	0	-
Mandarin Oranges,...	0	6.30	24.94	0
Red Tomato	0	4.61	3.90	0
Dried Coconut, Sh...	0	0.46	0.4	0
Almonds, Dry Roas...	0	9.35	0	0
Avocado, Californ...	0	26.28	3.73	0
Carrots	0	15.45	4.79	0
Peach	0	3.92	6.46	0
Strawberries	0	17.28	42.33	0
Long Grain White ...	0	222.25	0	0
Red Bell or Sweet...	0	10.72	113.24	0
White Granulated ...	0	0	0	0

Mushrooms, Boiled...	0	8.42	1.87	-
LOUIS RICH DELI T...	-	-	0	-
DEL MONTE Pineapp...	-	-	6	0
PEPPERIDGE FARM B...	-	-	0	-
KRAFT Singles Swi...	0.32	-	0	-
GENERAL MILLS KIX...	2.83	635.60	11.34	1.89
GENERAL FOODS INT...	-	-	0	-
PILLSBURY Cinnamo...	-	-	0	-
Totals:	**5.55 µg**	**1225.31 µg**	**365.25 mg**	**4.33 µg**

Item Name	Vit A (RAE) (µg)	alpha-T (mg)	Calcium (mg)	Iron (mg)
Romaine Lettuce, ...	1477.99	0.29	27.71	0.81
Low Calorie Blue ...	1.26	0.23	28.47	0.15
Red Bell or Sweet...	353.38	0.20	2.60	0.16
Tofu, Firm, with ...	0	-	91.85	0.82
Lowfat Milk, 1%	144.90	0.7	263.51	0.85
Broccoli, Chopped...	464.88	1.5	31.20	0.52
Onions, Chopped	0.29	0.11	10.56	0.9
Whole Wheat Bread...	0.45	0.46	40.50	1.85
Mandarin Oranges,...	320.69	0.34	8.81	0.46
Red Tomato	77.61	0.9	3.7	0.8
Dried Coconut, Sh...	0	0.6	0.87	0.11
Almonds, Dry Roas...	0.8	5.96	75.41	1.27
Avocado, Californ...	18.88	0.45	5.51	0.25
Carrots	2966.13	0.29	26.83	0.24
Peach	96.80	0.54	5.88	0.24
Strawberries	2.61	0.8	11.52	0.30
Long Grain White ...	0	0.6	33.25	1.97
Red Bell or Sweet...	565.41	0.32	4.17	0.25
White Granulated ...	0	0	0.3	0
Mushrooms, Boiled...	0	0.4	2.80	0.81
LOUIS RICH DELI T...	0	-	0	0.19
DEL MONTE Pineapp...	0	-	0	0.18
PEPPERIDGE FARM B...	0	-	0	0.24
KRAFT Singles Swi...	54.50	-	202.50	0
GENERAL MILLS KIX...	286.33	0.12	283.50	15.30
GENERAL FOODS INT...	0	-	0	0
PILLSBURY Cinnamo...	0	-	0	0.38
Totals:	**6832.91 µg**	**14.81 mg**	**1161.45 mg**	**29.5 mg**

Item Name	Magn (mg)	Potas (mg)	Zinc (mg)	Sodium (mg)
Romaine Lettuce, ...	11.76	207.47	0.19	6.71
Low Calorie Blue ...	2.24	1.60	0.7	384
Red Bell or Sweet...	4.46	78.59	0.9	0.74
Tofu, Firm, with ...	26.8	99.79	0.57	4.53
Lowfat Milk, 1%	26.84	290.35	2.12	122
Broccoli, Chopped...	16.37	228.53	0.35	31.97
Onions, Chopped	4.80	69.11	0.7	1.44
Whole Wheat Bread...	48.50	141.50	1.9	296
Mandarin Oranges,...	10.7	98.27	0.30	7.55
Red Tomato	3.38	72.87	0.5	1.53
Dried Coconut, Sh...	2.90	19.58	0.10	15.22
Almonds, Dry Roas...	81.8	211.49	1	96.10
Avocado, Californ...	12.29	214.97	0.28	3.39
Carrots	9.75	260.26	0.19	56.11
Peach	8.81	186.19	0.16	0

Item Name	Amount	Fol (DFE)	0% 25% 50% 75% 100%
GENERAL MILLS KIX...	2.0 oz.	636 µg	52%
Long Grain White ...	1.0 c.	222 µg	18%
Romaine Lettuce, ...	1.5 c.	114 µg	9%
Broccoli, Chopped...	0.5 c.	84 µg	7%
Avocado, Californ...	0.25 item	26 µg	2%
Whole Wheat Bread...	2.0 sl.	20 µg	2%
Tofu, Firm, with ...	2.0 oz.	19 µg	2%
Strawberries	0.5 c.	17 µg	1%
Carrots	0.67 c.	15 µg	1%
Lowfat Milk, 1%	1.0 c.	12 µg	1%
Red Bell or Sweet...	0.4 c.	11 µg	1%
Almonds, Dry Roas...	1.0 oz.	9 µg	1%
Onions, Chopped	0.3 c.	9 µg	1%
Mushrooms, Boiled...	0.3 c.	8 µg	1%
Red Bell or Sweet...	0.25 c.	7 µg	1%
Mandarin Oranges,...	0.5 c.	6 µg	1%
Red Tomato	0.25 item	5 µg	0%
Peach	1.0 item	4 µg	0%
Dried Coconut, Sh...	1.0 t.	0.46 µg	0%
White Granulated ...	1.0 t.	0 µg	0%
Low Calorie Blue ...	2.0 t.	* µg	--%
LOUIS RICH DELI T...	1.0 oz.	* µg	--%
DEL MONTE Pineapp...	0.25 c.	* µg	--%
PEPPERIDGE FARM B...	2.0 item	* µg	--%
KRAFT Singles Swi...	1.0 oz.	* µg	--%
GENERAL FOODS INT...	2.0 c.	* µg	--%
PILLSBURY Cinnamo...	0.5 item	* µg	--%
Totals:		1225 µg	

!	Nutrient (Minnie Mineral)	Intake	DRI	0% 50% 100%
	Energy			
	Calories	1370 kCal	2161.7 kCal	63%
	Carbohydrates	207 g	243 g - 351 g	
	Fat	32 g	48 g - 84 g	
	Protein	81 g	54 g - 189 g	
	Protein(g/kg/day)	81 g	51 g	159%
	Fat			
	Saturated Fat	10.23 g	*no rec*	
	Monounsaturated Fat	9.59 g *	*no rec*	
	Polyunsaturated Fat	7.01 g *	*no rec*	
	Cholesterol	160.44 mg	300 mg	53%
	Essential fatty acids			
	PFA 18:2, Linoleic	4.2 g *	12 g	35%
	PFA 18:3, Linolenic	0.85 g *	1.1 g	77%
	carbohydrates			
	Dietary Fiber, Total	25.15 g	25 g	101%
	Sugar, Total	95.8 g	*no rec*	
	Other			
	Water	0.98 L *	2.3 L	43%
	Alcohol	0 g	*no rec*	
	Vitamins			
	Thiamin (Vit B1)	1.72 mg *	1.1 mg	156%
	Riboflavin	2.54 mg *	1.1 mg	231%
	Niacin	33.67 mg *	14 mg	241%
	Vitamin B6	2.35 IU *	1.3 IU	181%
	Vitamin B12	11.01 µg *	2.4 µg	459%
	Folate (DFE)	117.9 µg *	400 µg	29%
	Vitamin C	79.62 mg	75 mg	106%
	Vitamin D (ug)	14.76 µg *	5 µg	295%
	Vitamin A(RAE)	4293.34 µg	699.9 µg	613%
	alpha-tocopherol (Vit E)	4.69 mg *	15.0 mg	31%
	Minerals			
	Calcium	1309.34 mg	1000 mg	131%
	Iron	19.09 mg	18 mg	106%
!	Magnesium	432.66 mg *	310 mg	140%
	Potassium	2576.28 mg *	4700 mg	55%
	Zinc	13.33 mg *	8 mg	167%
!	Sodium	2463.63 mg	1500 mg	164%

* This nutrient may be under-represented as not all foods in the selected report range have a value for this nutrient.

Macaroni, Cooked ...	0.11	0.5	0.90	0.2
Hard Boiled Egg	0.1	0.8	0.1	0.2
Sweet Pickle Reli...	0	0	0.3	0
Whole Wheat Bread...	0.15	0.10	1.94	0.9
Brownie, Small, R...	0.7	0.5	0.48	0
Light Tuna, Canne...	0.1	0.4	7.52	0.19
VLASIC Kosher Dil...	-	-	-	-
KRAFT Fat Free Ma...	-	-	-	-
POST BRAN FLAKES ...	0.53	0.60	7.8	0.70
OCEAN SPRAY CRAIS...	-	-	-	-
NATURAL OVENS Who...	0.44	0.50	8	0.60
Totals:	**4.53 mg**	**4.87 mg**	**35.39 mg**	**5.41 IU**

Item Name	Vit B12 (μg)	Fol (DFE) (μg)	Vit C (mg)	Vit D (ug) (μg)
Iceberg Lettuce L...	0	8.95	0.62	0
Pink Salmon, Soli...	2.49	8.50	0	8.84
Celery, Diced	0	2.70	0.23	0
Carrots	0	7.68	2.38	0
Nonfat, Skim or F...	0.71	9.18	1.83	1.83
Cantaloupe	0	-	58.77	0
Mozzarella Cheese...	0.98	4.25	0	-
Nonfat, Skim or F...	0.95	12.25	2.45	2.45
Mayonnaise-type S...	0.3	0.88	0	-
Red Tomato, Diced	0	6.75	5.71	0
Nectarine	0	6.90	7.45	0
Macaroni, Cooked ...	0	-	0	0
Hard Boiled Egg	0.18	7.48	0	0.22
Sweet Pickle Reli...	0	0.15	0.15	0
Whole Wheat Bread...	0	19.50	0	-
Brownie, Small, R...	0.1	20.44	0	-
Light Tuna, Canne...	1.69	2.26	0	-
VLASIC Kosher Dil...	0	-	0	0
KRAFT Fat Free Ma...	-	-	0	-
POST BRAN FLAKES ...	2.12	-	0	1.41
OCEAN SPRAY CRAIS...	-	-	0	0
NATURAL OVENS Who...	1.79	-	0	-
Totals:	**11.31 μg**	**117.87 μg**	**79.59 mg**	**14.75 μg**

Item Name	Vit A (RAE) (μg)	alpha-T (mg)	Calcium (mg)	Iron (mg)
Iceberg Lettuce L...	15.61	0.3	3.20	0.5
Pink Salmon, Soli...	9.78	0.61	120.70	0.47
Celery, Diced	10.20	0.2	3	0.1
Carrots	1474.21	0.14	13.33	0.12
Nonfat, Skim or F...	113.2	0.5	237.3	0.9
Cantaloupe	1814.18	-	20.39	0.33
Mozzarella Cheese...	66.61	0.15	310.85	0.10
Nonfat, Skim or F...	150.69	0.7	316.4	0.12
Mayonnaise-type S...	9.79	0.47	2.5	0.2
Red Tomato, Diced	113.57	0.13	4.50	0.12
Nectarine	138.82	0.98	8.27	0.38
Macaroni, Cooked ...	0	0.1	6	0.97
Hard Boiled Egg	30.18	0.14	8.50	0.20
Sweet Pickle Reli...	8.31	0	0.44	0.13
Whole Wheat Bread...	0.45	0.46	40.50	1.85
Brownie, Small, R...	5.93	0.47	8.23	0.63
Light Tuna, Canne...	9.79	0.24	6.23	0.86

VLASIC Kosher Dil...	0	-	0	0
KRAFT Fat Free Ma...	0	0.64	0	0
POST BRAN FLAKES ...	322.12	-	0	11.48
OCEAN SPRAY CRAIS...	0	-	0	0
NATURAL OVENS Who...	0	0	200	1.8
Totals:	**4293.44 µg**	**6.23 mg**	**1310.34 mg**	**21.26 mg**

Item Name	Magn (mg)	Potas (mg)	Zinc (mg)	Sodium (mg)
Iceberg Lettuce L...	1.28	24.31	0.2	1.43
Pink Salmon, Soli...	19.26	184.73	0.52	313.93
Celery, Diced	0.82	19.50	0	6
Carrots	4.85	129.35	0.9	27.89
Nonfat, Skim or F...	27.56	314.21	0.75	97.38
Cantaloupe	19.12	375.74	0.28	31.74
Mozzarella Cheese...	11.5	40.39	1.33	224.53
Nonfat, Skim or F...	36.75	418.95	1	129.85
Mayonnaise-type S...	0.29	1.32	0.2	104.51
Red Tomato, Diced	4.94	106.65	0.7	2.25
Nectarine	12.42	277.38	0.23	0
Macaroni, Cooked ...	14.17	48	0.41	0.75
Hard Boiled Egg	1.70	21.42	0.17	21.8
Sweet Pickle Reli...	0.75	3.75	0.2	121.65
Whole Wheat Bread...	48.50	141.50	1.9	296
Brownie, Small, R...	8.80	42.31	0.20	88.60
Light Tuna, Canne...	15.30	134.37	0.43	191.64
VLASIC Kosher Dil...	-	-	-	52.50
KRAFT Fat Free Ma...	-	20	-	240
POST BRAN FLAKES ...	85.5	269.32	2.12	311.85
OCEAN SPRAY CRAIS...	-	3.3	-	0
NATURAL OVENS Who...	120	0	4.50	200
Totals:	**433.51 mg**	**2576.50 mg**	**16.4 mg**	**2464.30 mg**

Item Name	Amount	Sodium		0% 25% 50% 75% 100%
Pink Salmon, Soli...	2.0 oz.	314 mg	✩	13%
POST BRAN FLAKES ...	1.5 oz.	312 mg	✩	13%
Whole Wheat Bread...	2.0 sl.	296 mg	✩	12%
KRAFT Fat Free Ma...	2.0 t.	240 mg		10%
Mozzarella Cheese...	1.5 oz.	225 mg		9%
NATURAL OVENS Who...	1.0 item	200 mg		8%
Light Tuna, Canne...	2.0 oz.	192 mg		8%
Nonfat, Skim or F...	1.0 c.	130 mg		5%
Sweet Pickle Reli...	1.0 t.	122 mg		5%
Mayonnaise-type S...	1.0 t.	105 mg		4%
Nonfat, Skim or F...	0.75 c.	97 mg		4%
Brownie, Small, R...	1.0 item	89 mg		4%
VLASIC Kosher Dil...	0.25 oz.	53 mg		2%
Cantaloupe	0.25 item	32 mg		1%
Carrots	0.333 c.	28 mg		1%
Hard Boiled Egg	2.0 t.	21 mg		1%
Celery, Diced	1.0 t.	6 mg		0%
Red Tomato, Diced	0.25 c.	2 mg		0%
Iceberg Lettuce L...	2.0 pc.	1 mg		0%
Macaroni, Cooked ...	0.75 c.	0.75 mg		0%
Nectarine	1.0 svg.	0 mg		0%
OCEAN SPRAY CRAIS...	1.0 t.	0 mg		0%
Totals:		2464 mg		

Item Name	Amount	Calcium		0% 25% 50% 75% 100%
Nonfat, Skim or F...	1.0 c.	316 mg	★	24%
Mozzarella Cheese...	1.5 oz.	311 mg	★	24%
Nonfat, Skim or F...	0.75 c.	237 mg	★	18%
NATURAL OVENS Who...	1.0 item	200 mg	★	15%
Pink Salmon, Soli...	2.0 oz.	121 mg	★	9%
Whole Wheat Bread...	2.0 sl.	41 mg		3%
Cantaloupe	0.25 item	20 mg		2%
Carrots	0.333 c.	13 mg		1%
Hard Boiled Egg	2.0 t.	9 mg		1%
Nectarine	1.0 svg.	8 mg		1%
Brownie, Small, R...	1.0 item	8 mg		1%
Light Tuna, Canne...	2.0 oz.	6 mg		0%
Macaroni, Cooked ...	0.75 c.	6 mg		0%
Red Tomato, Diced	0.25 c.	5 mg		0%
Iceberg Lettuce L...	2.0 pc.	3 mg		0%
Celery, Diced	1.0 t.	3 mg		0%
Mayonnaise-type S...	1.0 t.	2 mg		0%
Sweet Pickle Reli...	1.0 t.	0.44 mg		0%
VLASIC Kosher Dil...	0.25 oz.	0 mg		0%
KRAFT Fat Free Ma...	2.0 t.	0 mg		0%
POST BRAN FLAKES ...	1.5 oz.	0 mg		0%
OCEAN SPRAY CRAIS...	1.0 t.	0 mg		0%
Totals:		1309 mg		

Profile

Profile Name	Minnie Mineral
Height	5 ft. 5 inches
Weight	140 lb.
Age	29 years
BMI	23

DRI Goals

Nutrient	DRI	
Energy Nutrients		
Calories (kcal)	2162 kcal	
Carbohydrates	243 - 351 g	*45%-65% of kilocalories*
Fat	48 - 84 g	*20%-35% of kilocalories*
Protein	54 - 189 g	*10%-35% of kilocalories*
Protein (g/kg)	51 g	*Daily requirement based on grams per kilogram of body weight*
Fat		
Saturated Fat	24	*less than 10% of calories recommended*
Monounsaturated Fat	-	*No recommendation*
Polyunsaturated Fat	-	*No recommendation*
Cholesterol	300 mg	*less than 300mg recommended*
Essential fatty acids		
PFA 18:2, Linoleic	12 g	
PFA 18:3, Linolenic	1.1 g	
Carbohydrates		
Dietary Fiber, Total	25 g	
Sugar, Total	-	*No recommendation*
Other		
Water	2.3 L	
Alcohol	-	*No recommendation*
Vitamins		
Thiamin (Vit B1)	1.1 mg	
Riboflavin	1.1 mg	
Niacin	14 mg	
Vitamin B6	1.3 IU	
Vitamin B12	2.4 µg	
Folate (DFE)	400 µg	
Vitamin C	75 mg	
Vitamin D (ug)	5 µg	
Vitamin A(RAE)	699.9 µg	
alpha-tocopherol (Vit E)	15 mg	
Minerals		
Calcium	1000 mg	*DRI Adequate Intake*
Iron	18 mg	
Magnesium	310 mg	
Potassium	4700 mg	*DRI Adequate Intake*
Zinc	8 mg	
Sodium	1500 mg	*DRI Adequate Intake*

Profile

Profile Name	Minnie Mineral, pregnant
Height	5 ft. 5 inches
Weight	140 lb.
Age	29 years
BMI	23

DRI Goals

Nutrient	DRI	
Energy Nutrients		
Calories (kcal)	2502 kcal	
Carbohydrates	281 - 407 g	*45%-65% of kilocalories*
Fat	56 - 97 g	*20%-35% of kilocalories*
Protein	63 - 219 g	*10%-35% of kilocalories*
Protein (g/kg)	70 g	*Daily requirement based on grams per kilogram of body weight*
Fat		
Saturated Fat	28	*less than 10% of calories recommended*
Monounsaturated Fat	-	*No recommendation*
Polyunsaturated Fat	-	*No recommendation*
Cholesterol	300 mg	*less than 300mg recommended*
Essential fatty acids		
PFA 18:2, Linoleic	13 g	
PFA 18:3, Linolenic	1.4 g	
carbohydrates		
Dietary Fiber, Total	28 g	
Sugar, Total	-	*No recommendation*
Other		
Water	3 L	
Alcohol	-	*No recommendation*
Vitamins		
Thiamin (Vit B1)	1.4 mg	
Riboflavin	1.4 mg	
Niacin	18 mg	
Vitamin B6	1.9 IU	
Vitamin B12	2.6 µg	
Folate (DFE)	600 µg	
Vitamin C	85 mg	
Vitamin D (ug)	5 µg	
Vitamin A(RAE)	749.9 µg	
alpha-tocopherol (Vit E)	15 mg	
Minerals		
Calcium	1000 mg	*DRI Adequate Intake*
Iron	27 mg	
Magnesium	350 mg	
Potassium	4700 mg	*DRI Adequate Intake*
Zinc	11 mg	
Sodium	1500 mg	*DRI Adequate Intake*

Exercise #1 – 7-Day Diet Analysis Summary

Use the *Diet Analysis Plus* Intake vs. Goals and Macronutrient Ranges reports (print one for each day, plus one for the average of all 7 days) to complete the following table. Select one vitamin and one mineral to include in this summary (use the same vitamin and mineral each day). Include the name of the vitamin and mineral and units of measure (mg or mcg). Place the "value" in the top half of each cell and the "goal %" in the bottom half of each cell. In the space below each day, include brief remarks/notes about this day's intake etc, such as food items, activities, etc. The objective of this summary is to start to look at your intake of nutrients in order to begin to see patterns, strengths and weakness in your diet (to improve or maintain your health and reduce the risk of disease).

	Total kcal	Carbohydrate	Fat	Protein	Vitamin: ______	Mineral: ______
Day 1 value						
Day 1 %						
Day 2 value						
Day 2 %						
Day 3 value						
Day 3 %						
Day 4 value						
Day 4 %						
Day 5 value						
Day 5 %						
Day 6 value						
Day 6 %						
Day 7 value						
Day 7 %						
Average val.						
Average %						

Using your Profile DRI Goals report, list your recommended intake of:

Kcal	
Carbohydrate (g)	
Dietary Fiber (g)	
Fat (g)	
Saturated fat (g)	
Cholesterol (mg)	
Protein (g)	
Vitamin (selected) – name and units	
Mineral (selected) – name and units	

What are your AMDR values as % of kcal for each of the energy nutrients?

Energy Nutrient	**AMDR**	**Your Selection (to total 100%)**
Carbohydrate	45-65%	
Fat	20-35%	
Protein	10-35%	

EXERCISE #2 – EVALUATION OF YOUR 7-DAY PROTEIN INTAKE USING *DIET ANALYSIS PLUS*

1. **Refer to your Profile DRI Goals** (take these numbers directly off your printouts):

   What is your RDA for kcal? _____ kcal per day

   What is your goal for carbohydrate _____ grams per day

   What is your goal for fat? _____ grams per day

   What is your goal for protein? _____ grams per day

2. Calculate your DRI recommended intake for protein based on your healthy body weight. **Place your answer here**: ________

3. What AMDR percentage have you selected for your protein intake? ______________ **(fill in)**

   Using the kcal goal listed on your Profile DRI Goals printout and your AMDR percentage, determine a goal (in grams) for your protein intake. Compare your answer to the RDA calculated in question 2 above and the goal for protein listed on your Profile DRI Goals report.

4. Using your *Diet Analysis Plus* Intake vs. Goals reports (print one for each day, plus one for the average of all 7 days), fill in the table below.

	Total kcal	Protein	Total fat	Sat. fat	Mono. fat	Cholesterol
Day 1 value						
Day 1 %						
Day 2 value						
Day 2 %						
Day 3 value						
Day 3 %						
Day 4 value						
Day 4 %						
Day 5 value						
Day 5 %						
Day 6 value						
Day 6 %						
Day 7 value						
Day 7 %						
Average val.						
Average %						

5. Using the data from this table, what general patterns do you see regarding your intake? **Attach your Intake Spreadsheet to your assignment.**

   - Looking at your values in this table, do you find that when your intake of kcal increases, your intakes of protein, fat, saturated fat and cholesterol also increase?

   - What day did you have the most or the least protein (g/day)? Consider what foods you ate on those days.

   - Are you surprised at your level of protein intake? Why or why not?

6. Do you have any days where the nutrient density ratio for protein is less than 1? (To calculate the nutrient density ratio, divide % nutrient DRI goal by % kcal DRI goal supplied.) If your answer is yes, what day(s)? **Place your answer here and show the calculation of the nutrient density ratio.**

7. For the next calculation, use the day your protein intake was the **highest**. Refer to the **Intake Spreadsheet** for this day and put an "A" by animal sources of protein. You may need to "guesstimate" as some items may contain both animal and plant sources of protein (cheese pizza for example). Calculate the ratio of animal protein to total protein in your diet (grams of animal protein divided by total grams of protein). **Include your spreadsheet (with your assignment).**

8. Using the day where your protein intake was the highest, identify your significant sources of animal (>4 grams/serving) and your significant sources of plant (>2 grams/serving) protein. Place your answers below. Specify the serving size consumed and the grams of protein provided.

Animal protein sources:	
Food and serving size consumed	Grams of protein provided

Plant protein sources:	
Food and serving size consumed	Grams of protein provided

- This information can be found on your **Intake Spreadsheet** or you can use the **Source Analysis** report from *Diet Analysis Plus*, choosing protein as the nutrient to analyze. Keep in mind that some foods supply both plant protein and animal protein (a sandwich for example); you will need to estimate amounts accordingly.
- **Attach the printout used for this question to your assignment.**

9. What percentage of your kcal are supplied by protein, carbohydrate and fat? Compare to your AMDR goals. This is listed on the **Macronutrient Ranges** report.

	% kcal from protein	% kcal from carbohydrate	% kcal from fat
Day 1			
Day 2			
Day 3			
Day 4			
Day 5			
Day 6			
Day 7			
7-day average			
	AMDR for protein	AMDR for carbohydrate	AMDR for fat
	10-35%	45-65%	20-35%

EXERCISE #3 – 7-DAY VITAMIN AND MINERAL EVALUATION

1. Using your *Diet Analysis Plus* reports, complete the following tables. The information is found on your Intake vs. Goals reports (one for each day, plus one for the average of all 7 days). To calculate nutrient density, divide the % of the RDA (or AI) of the vitamin supplied by the % of the kcal goal supplied (labeled below as column 3 divided by column 4) and place your answer in column 5 (**keep two decimal places in your nutrient density ratio**). A nutrient density ratio greater than 1 means that your vitamin (or mineral) intake is proportional to your kcal intake. If the ratio is less than 1, your intake of the vitamin is low for your intake of kcal.

Vitamin A Your RDA for vitamin A = ________ μg RAE/day (Retinol Activity Equivalents)

	Column 2 vitamin A intake μg/day	Column 3 % vitamin A RDA supplied	Column 4 % goal kcal supplied	Column 5 nutrient density ratio
Day 1				
Day 2				
Day 3				
Day 4				
Day 5				
Day 6				
Day 7				
7-day average				

Vitamin C Your RDA for vitamin C = ________ mg/day

	vitamin C intake mg/day	% vitamin C RDA supplied	% goal kcal supplied	nutrient density ratio
Day 1				
Day 2				
Day 3				
Day 4				
Day 5				
Day 6				
Day 7				
7-day average				

B Vitamin Select and circle one of the following: **thiamin, riboflavin** or **niacin**.

Your RDA for the selected B vitamin = ________ mg/day

	B vitamin intake mg/day	% vitamin RDA supplied	% goal kcal supplied	nutrient density ratio
Day 1				
Day 2				
Day 3				
Day 4				
Day 5				
Day 6				
Day 7				
7-day average				

2. **Food Sources of Vitamins.** For each vitamin, select three days as follows: your highest intake day, your lowest intake day, and one that comes closest to your 7-day average. Complete one table for each of your three days (see below).

- Identify the "day" by putting the date of the intake (see heading for each table).
- List (by name) your three highest animal and three highest plant sources for each vitamin; put your sources in order from most to least (you can use *Diet Analysis Plus* to sort your foods by single nutrients; the Source Analysis report lists the foods in order of amount supplied from most to least).
- Specify the serving size consumed and the amount of the vitamin supplied. Some vitamins are mostly supplied by animal foods while others are primarily supplied by plant foods. Refer to column headings when filling in this information.
- After you have entered your three highest sources, divide the amount supplied by your RDA or AI and express as a percentage (see table below). Repeat for each vitamin listed.
- Tables follow (3 tables for each vitamin).
- The information in the tables from Questions 1 and 2 will be useful for analyzing your vitamin intake – see Question 3.

Vitamin A Highest intake day ______________ (list date of intake)

animal food	serving size	amount supplied µg RAE	% RDA	plant food	serving size	amount supplied µg RAE	% RDA

Vitamin A Lowest intake day ______________ (list date of intake)

animal food	serving size	amount supplied µg RAE	% RDA	plant food	serving size	amount supplied µg RAE	% RDA

Vitamin A Day closest to the 7-day average intake ______________ (list date of intake)

animal food	serving size	amount supplied µg RAE	% RDA	plant food	serving size	amount supplied µg RAE	% RDA

When considering food sources, use your RDA or AI as a reference point. Foods with only 1-2% of your RDA are not significant sources. A significant source supplies 25% of the RDA; an excellent source supplies 75%; and a fair source supplies 10%. When you look at your three highest sources, make a generalization about this day: Is it high or low? Do you need to include other foods to meet your RDA? Is most of your vitamin intake from animal foods or plant foods or both?

Vitamin C Highest intake day ______________ (list date of intake)

animal food	serving size	amount supplied mg	% RDA	plant food	serving size	amount supplied mg	% RDA

Vitamin C Lowest intake day ______________ (list date of intake)

animal food	serving size	amount supplied mg	% RDA	plant food	serving size	amount supplied mg	% RDA

Vitamin C Day closest to the 7-day average intake ________________ (list date of intake)

animal food	serving size	amount supplied mg	% RDA	plant food	serving size	amount supplied mg	% RDA

Select one of the following B-vitamins to complete these tables (thiamin, riboflavin, niacin, B_6, folic acid, or B_{12}). List the RDA for this vitamin first (for reference):

B vitamin selected ________________________________ RDA = ________ (mg or μg) Circle

B vitamin Highest intake day ________________ (list date of intake)

animal food	serving size	amount supplied mg / μg	% RDA	plant food	serving size	amount supplied mg / μg	% RDA

B vitamin Lowest intake day ________________ (list date of intake)

animal food	serving size	amount supplied mg / μg	% RDA	plant food	serving size	amount supplied mg / μg	% RDA

B vitamin Day closest to the 7-day average intake ________________ (list date of intake)

animal food	serving size	amount supplied mg / μg	% RDA	plant food	serving size	amount supplied mg / μg	% RDA

3. **Analyzing food sources of vitamins using food groups – Summary and Discussion.** Analyze your intake of vitamins using the Food Pyramid Analysis report. Suggest ways to maximize intake of vitamins from your usual food choices. Use your *Diet Analysis Plus* reports as appropriate to support your answer. **Suggestion**: Organize your discussion by food groups and evaluate the foods in each group as sources of vitamins. Length: 2-3 pages (word-processed; double-spaced; 12 point font).

4. Complete the following tables with information on your mineral intake using your *Diet Analysis Plus* reports. The information is found on your Intake vs. Goals reports (one for each day, plus one for the average of all 7 days). To calculate nutrient density, divide the % of the RDA (or AI) of the vitamin supplied by the % of the kcal goal supplied (labeled below as column 3 divided by column 4) and place your answer in column 5 (**keep two decimal places in your nutrient density ratio**). A nutrient density ratio greater than 1 means that your vitamin (or mineral) intake is proportional to your kcal intake. If the ratio is less than 1, your intake of the vitamin is low for your intake of kcal.

<u>**Calcium**</u> Your AI for calcium = ________ mg/day

	Column 2 calcium intake mg/day	Column 3 % calcium AI supplied	Column 4 % goal kcal supplied	Column 5 nutrient density ratio
Day 1				
Day 2				
Day 3				
Day 4				
Day 5				
Day 6				
Day 7				
7-day average				

<u>**Iron**</u> Your RDA for iron = ________ mg/day

	iron intake mg/day	% iron RDA supplied	% goal kcal supplied	nutrient density ratio
Day 1				
Day 2				
Day 3				
Day 4				
Day 5				
Day 6				
Day 7				
7-day average				

The calculations in the table for sodium (on the following page) are slightly different. In Column 2, place your intake of sodium in mg; in column 3, place the % of your AI; in column 4, place your kcal intake (the amount, not the % goal); and in column 5, divide your intake in mg (column 2) by kcal (column 4) to get mg of sodium consumed per kcal. A general guideline is to consume no more than 1-2 mg sodium per kcal.

Sodium Your AI for sodium = ________ mg/day

	Column 2 sodium intake mg/day	Column 3 % sodium AI supplied	Column 4 kcal intake	Column 5 mg sodium/kcal
Day 1				
Day 2				
Day 3				
Day 4				
Day 5				
Day 6				
Day 7				
7-day average				

5. **Food Sources of Minerals (calcium, iron, sodium and your choice).** For each mineral, select three days as follows: your highest intake day, your lowest intake day and one that comes closest to your 7-day average. Complete one table for each of your three days (see below).

- Identify the "day" by putting the date of the intake (see heading for each table).
- List (by name) your three highest animal and three highest plant sources for each mineral; put your sources in order from most to least (you can use *Diet Analysis Plus* to sort your foods by single nutrients; the Source Analysis report lists the foods in order of amount supplied from most to least).
- Specify the serving size consumed and the amount of the mineral supplied. Some minerals are mostly supplied by animal foods while others are primarily supplied by plant foods; some minerals are provided by both animal and plant foods. Refer to column headings when filling in this information.
- After you have entered your three highest sources, divide the amount supplied by your RDA or AI and express as a percentage (see table below). Repeat for each mineral listed below.
- Tables follow (3 tables for each mineral).

Calcium Highest intake day ________________ (list date of intake)

animal food	serving size	amount supplied mg	% AI	plant food	serving size	amount supplied mg	% AI

Calcium Lowest intake day ________________ (list date of intake)

animal food	serving size	amount supplied mg	% AI	plant food	serving size	amount supplied mg	% AI

Calcium Day closest to the 7-day average intake ________________ (list date of intake)

animal food	serving size	amount supplied mg	% AI	plant food	serving size	amount supplied mg	% AI

When considering food sources, use your RDA or AI as a reference point. Foods with only 1-2% of your RDA are not significant sources. A significant source supplies 25% of the RDA; an excellent source supplies 75%; and a fair source supplies 10%. When you look at your three highest sources, make a generalization about this day: Is it high or low? Do you need to include other foods to meet your recommended intake? Is most of your mineral intake from animal foods or plant foods or both?

Iron Highest intake day ________________ (list date of intake)

animal food	serving size	amount supplied mg	% RDA	plant food	serving size	amount supplied mg	% RDA

Iron Lowest intake day ________________ (list date of intake)

animal food	serving size	amount supplied mg	% RDA	plant food	serving size	amount supplied mg	% RDA

Iron Day closest to the 7-day average intake ________________ (list date of intake)

animal food	serving size	amount supplied mg	% RDA	plant food	serving size	amount supplied mg	% RDA

Sodium Highest intake day ________________ (list date of intake)

animal food	serving size	amount supplied mg	% AI	plant food	serving size	amount supplied mg	% AI

Sodium Lowest intake day ________________ (list date of intake)

animal food	serving size	amount supplied mg	% AI	plant food	serving size	amount supplied mg	% AI

Sodium Day closest to the 7-day average intake ________________ (list date of intake)

animal food	serving size	amount supplied mg	% AI	plant food	serving size	amount supplied mg	% AI

Select one additional mineral to evaluate (your choice):

Mineral selected ________________________________ RDA or AI = ________ (mg or μg) Circle

Selected mineral Highest intake day ______________ (list date of intake)

animal food	serving size	amount supplied mg / μg	% RDA or AI	plant food	serving size	amount supplied mg / μg	% RDA or AI

Selected mineral Lowest intake day ______________ (list date of intake)

animal food	serving size	amount supplied mg / μg	% RDA or AI	plant food	serving size	amount supplied mg / μg	% RDA or AI

Selected mineral Day closest to the 7-day average intake ______________ (list date of intake)

animal food	serving size	amount supplied mg / μg	% RDA or AI	plant food	serving size	amount supplied mg / μg	% RDA or AI

6. **Written evaluation of mineral intake.** Evaluate your mineral intake. Focus your discussion on your diet, your RDA/DRI, and the foods you eat (or should eat more or less of). Support your statements with information from the tables completed above (questions 4 and 5) as well as information from your *Diet Analysis Plus* printouts. Recommend dietary changes to improve your intake of minerals. Support your evaluation with specific examples. Length: 2-3 pages (word-processed; double-spaced; 12 point font). Attach.

EXERCISE #4 – EVALUATION OF 3-DAY VITAMIN INTAKE USING *DIET ANALYSIS PLUS*

1. Using your *Diet Analysis Plus* reports, complete the following tables.

Vitamin A Your RDA for Vitamin A = ________ μg RAE

	μg RAE supplied per day	% vitamin A RDA supplied	% goal kcal supplied	nutrient density ratio
Day 1				
Day 2				
Day 3				
3-day average				

To calculate the nutrient density ratio, divide % nutrient RDA supplied by % of goal kcal supplied. A nutrient density ratio greater than 1 means that your vitamin A intake and your kcal intake are proportional. For vitamin A, it is often possible to supply almost all of it from a single food (which is also often low in calories).

Vitamin C Your RDA for Vitamin C = ________ mg

	milligrams supplied per day	% vitamin C RDA supplied	% goal kcal supplied	nutrient density ratio
Day 1				
Day 2				
Day 3				
3-day average				

To calculate the nutrient density ratio, divide % nutrient RDA supplied by % of goal kcal supplied. A nutrient density ratio greater than 1 means that your vitamin C intake and your kcal intake are proportional. For vitamin C, it is often possible to supply almost all of it from a single food.

B Vitamin: thiamin, riboflavin or **niacin** (choose and circle one B vitamin)

Your RDA for this B vitamin = ________ mg

	milligrams supplied per day	% vitamin RDA supplied	% goal kcal supplied	nutrient density ratio
Day 1				
Day 2				
Day 3				
3-day average				

To calculate the nutrient density ratio, divide % nutrient RDA supplied by % of goal kcal supplied. A ratio greater than 1 means that your vitamin intake and your kcal intake are proportional. For most B-vitamins, it usually is not possible to meet your nutrient standards when kcaloric intake is inadequate.

2. For each vitamin select one of your three days to analyze. List your **five highest** animal and plant sources for each vitamin; put your sources **in order from most to least** (you can use *Diet Analysis Plus* to sort your foods by single nutrients; the printout lists the foods in order of amount supplied from

most to least). Specify the serving size consumed and the amount of the vitamin supplied. Some vitamins are mostly supplied by animal foods while others are primarily supplied by plant foods. **After your have entered your five highest sources, divide the amount supplied by your RDA or AI and express as a percentage. Enter this percentage in the table. Repeat for each vitamin listed below.**

Vitamin A **Day selected:** 1 2 3 (circle) RDA for Vitamin A = ________ µg RAE

animal food	serving size	amount supplied µg RAE	% RDA	plant food	serving size	amount supplied µg RAE	% RDA

Vitamin C **Day selected:** 1 2 3 (circle) RDA for Vitamin C = ________ mg

animal food	serving size	amount supplied mg	% RDA	plant food	serving size	amount supplied mg	% RDA

B Vitamin: ________________ (identify) **Day selected:** 1 2 3 RDA for B vitamin = ________ mg

animal food	serving size	amount supplied mg	% RDA	plant food	serving size	amount supplied mg	% RDA

3. Using the tables from question 2 and your text book, compare your food sources of vitamins A, C, and either thiamin, riboflavin or niacin to the significant food sources. For reference, remember that to be considered as a significant source of a nutrient, the food must supply 25% of the RDA in a serving (an excellent source supplies 75% of the RDA and a fair source supplies 10%). Discuss each vitamin in a separate paragraph. Word-process your answer. Suggested length: 2 pages double-spaced.

EXERCISE #5 – EVALUATION OF 3-DAY MINERAL INTAKE USING *DIET ANALYSIS PLUS*

1. Use your *Diet Analysis Plus* reports to complete the following tables. This information can be found on your Intake vs. Goals reports for each day and for the average of the 3 days.

Calcium Your AI for calcium = ________ mg

	mg supplied per day	% calcium AI supplied	% goal kcal supplied	nutrient density ratio
Day 1				
Day 2				
Day 3				
3-day average				

To calculate the nutrient density ratio, divide % nutrient AI or RDA supplied by % of goal kcal supplied. A nutrient density ratio greater than 1 means that your calcium intake and your kcal intake are proportional. For calcium, it is generally possible to supply almost all of your RDA from two or three nutrient dense foods.

Iron Your RDA for iron = ________ mg

	mg supplied per day	% iron RDA supplied	% goal kcal supplied	nutrient density ratio
Day 1				
Day 2				
Day 3				
3-day average				

To calculate the nutrient density ratio, divide % nutrient RDA supplied by % of goal kcal supplied. A ratio greater than 1 means that your iron intake and your kcal intake are proportional. For iron, it is difficult to meet your RDA without consuming adequate kcal.

Sodium: There are several guidelines or standards for evaluating sodium intakes.

Daily Value = ________ mg/day AI = ________ mg/day

	mg per day	kcal per day	mg sodium per kcal ratio [a]	% DV [b]	% AI [c]
Day 1					
Day 2					
Day 3					
3-day average					

[a] To calculate the milligram sodium per kcal ratio, divide milligrams sodium per day by kcal per day.
[b] To calculate % DV divide milligrams sodium per day by the DV and express as a percentage.
[c] To calculate the % AI divide milligrams sodium per day by the AI (for your age and gender) and express as a percentage.

2. Iron availability is calculated for each meal or snack. Select one meal from your three-day intake to use for this calculation. Use a meal with at least five foods. You will need to calculate the values for heme and non-heme iron. Heme iron is only found in animal flesh (MFP); 40% of the total iron in MFP is heme iron. Non-heme iron is found in all plant foods (nuts, seeds, fruits, vegetables, grains, legumes), milk and eggs plus the remaining 60% iron in MFP. A quick way to calculate non-heme iron is to first determine the amount of heme iron in a meal and then subtract this value from the total iron in the meal.

Food	Serving size	Total iron (mg)	Heme iron (mg)	Non-heme iron (mg)	Vitamin C (mg)
Totals:	---				

Using **your values from the above table**, calculate absorbable iron.

First, multiply your milligrams of heme iron by 25% and your milligrams of non-heme iron by 17% to estimate milligrams of each type of iron absorbed:

- ________ mg heme iron x **0.23** = ________ mg heme iron absorbed
- ________ mg non-heme iron x **0.17** = ________ mg non-heme iron absorbed

Note: Non-heme iron may be lower than 17% if factors enhancing non-heme iron absorption are absent (vitamin C and MFP factor, for example).

- Add the ________ mg heme iron absorbed and the ________ mg non-heme iron absorbed to get the **total**: ________ mg iron absorbed.
- Divide the **total** mg of iron absorbed (from above) by the amount of total iron in this meal and express as a percentage: ________%. Compare to the average percentage of iron absorbed from a mixed diet.

3. For each mineral (calcium, iron, and sodium) select one of your three days to analyze (you may use different days for each mineral).
   - List your **five highest** animal and plant sources for each mineral; put your sources **in order from most to least** (you can use *Diet Analysis Plus* to sort your foods by single nutrients; the Source Analysis report lists the foods in the order of the amount supplied from most to least).
   - Specify the serving size consumed and the amount of the mineral supplied. Some minerals are mostly supplied by animal foods while others are primarily supplied by plant foods.
   - After you have entered your five highest sources, divide the amount supplied by your RDA or AI and express as a percentage. Enter this percentage in the table. Repeat for each mineral.

Calcium **Day selected:** 1 2 3 (circle) AI = ________ mg

animal food	serving size	amount supplied mg	% AI	plant food	serving size	amount supplied mg	% AI

Iron **Day selected:** 1 2 3 (circle) RDA = ________ mg

animal food	serving size	amount supplied mg	% RDA	plant food	serving size	amount supplied mg	% RDA

Sodium **Day selected:** 1 2 3 (circle) AI = ________ mg

animal food	serving size	amount supplied mg	% RDA	plant food	serving size	amount supplied mg	% RDA

Select a mineral of your choice and identify below. Use appropriate units to fill in the RDA/AI.

Mineral: ____________________ **Day selected:** 1 2 3 (circle) RDA/AI = __________

animal food	serving size	amount supplied	% RDA or AI	plant food	serving size	amount supplied	% RDA or AI

4. Write summary of your mineral evaluation. Focus your discussion on your diet, your RDA/DRI, and the foods you eat (or should eat more or less of). Support your statements with information from the tables and questions above as well as your computer printouts. Recommend dietary changes to improve your intake of minerals. Word-process your summary. Suggested length: 2-3 pages double-spaced.

DIET ANALYSIS PLUS ONLINE ASSIGNMENT – PART A
Three-Day Dietary Intake

Web site: http://daplus7.wadsworth.com

The following is the screen that will appear. In order to register, you will need a registration code. This will be provided with the purchase of *Diet Analysis Plus Online* pin code card.

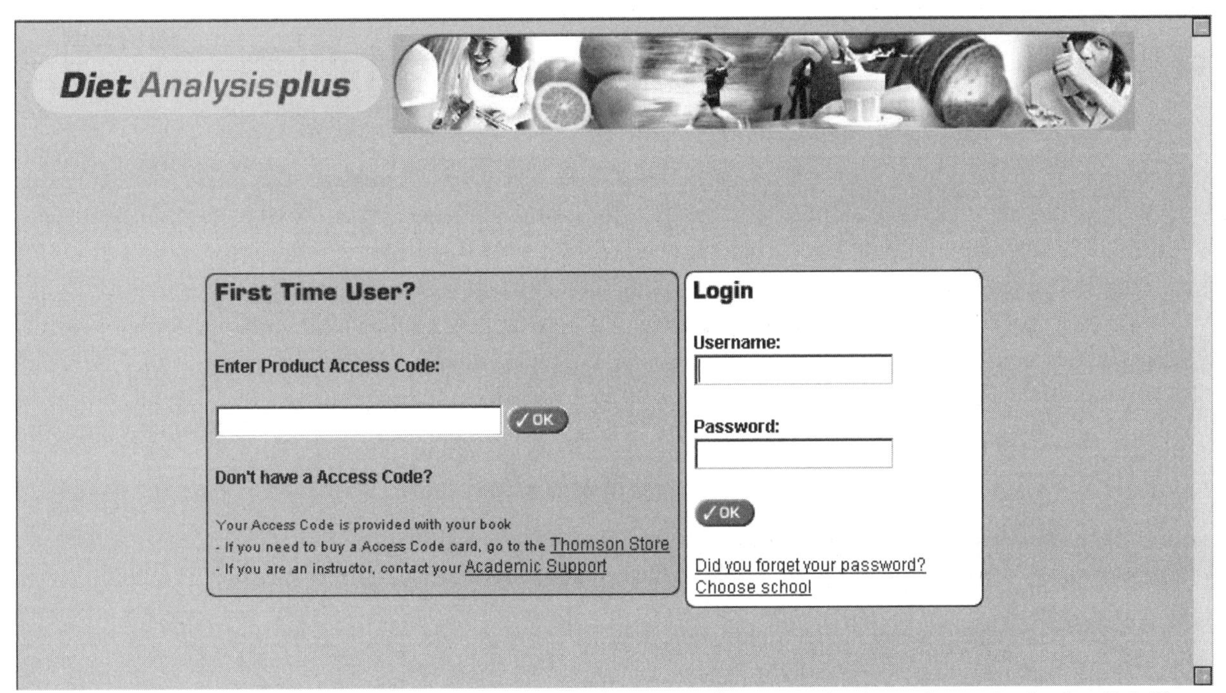

Once you have registered and have a username and password you can login to the web site for *DA Plus Online*. Write down your username and password (they are case sensitive) and keep this information in a safe and accessible place.

Once you are logged in, you will see the "Welcome to DA PLUS" screen. Click on **Profile** and enter the information that is requested:

- Name
- Age
- Sex
- Height and weight
- Whether you are a smoker or a vegetarian
- Activity level (If you click the **How Active am I?** button to the right, the definitions of each activity level will appear. You can also click on the **Typical Activities** button from this pop-up box to determine your activity level by entering your activities for a typical day.)

Your personal recommendations will be calculated based on your entries. You can edit your profile as you need to. To do so, click on the **Home** button in the navigation bar at the top of the screen, then **Profile**, then **Edit**.

The program is relatively simple to use and should be accessible to students of varying degrees of computer literacy. The following instructions are provided as additional information.

You are now ready to enter your food intake. Click on **Track Diet**. You can enter up to 365 days; however, for this assignment, only enter 3 days.

You will need to enter all foods and beverages (including alcohol and water). Try to include at least one weekend day in the three-day record. The days do not have to be consecutive. You can record the food you ate on a Tuesday, Thursday, and Sunday, for example. Record what you ate and the amount, which includes the quantity and unit of measure.

When you are ready to start entering your food, type in the name of the food (for example, Harmony cereal) and hit enter. The database will be searched for that food item. If nothing appears in the Food List box, and you see "0 items" in the blue title bar, no matches were found. If you just type in "cereal" (more generic) more food items will be found. They are displayed in groups of 30. You will need to advance to the next screen to see more food items. Select a cereal which is closest to the one you ate.

Example of how to enter a food item. Note—It is important to spell correctly; otherwise the search may find "0 items" as mentioned above.

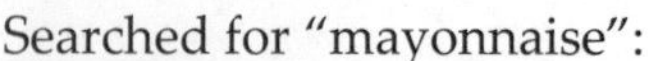
Searched for "mayonnaise":

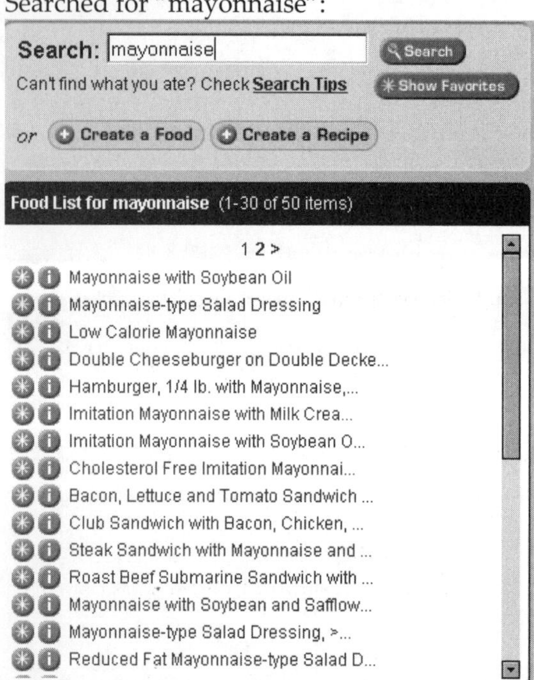

If you want to see the nutrition information about the food item, click on the **i** button to the left of the food and another screen will come up. This information may help you decide how to enter the amount of the food you ate. The computer automatically calculates the nutrition information for the amount you will enter.

To add the food item to your food list, select the item, enter the quantity under "How much did you eat?" and serving size under "Of what serving size?" in the pop-up box. Quantities may be entered as whole numbers or decimals (such as .5, if you ate half of a banana). You can click on "How big is a serving?" for some hints on estimating serving sizes. Then click the **Breakfast**, **Lunch**, **Dinner** or **Snack** button to indicate when you ate or drank that item. The food item will be added to your food list for that day; you can delete items by clicking the **X** button to the left of the item.

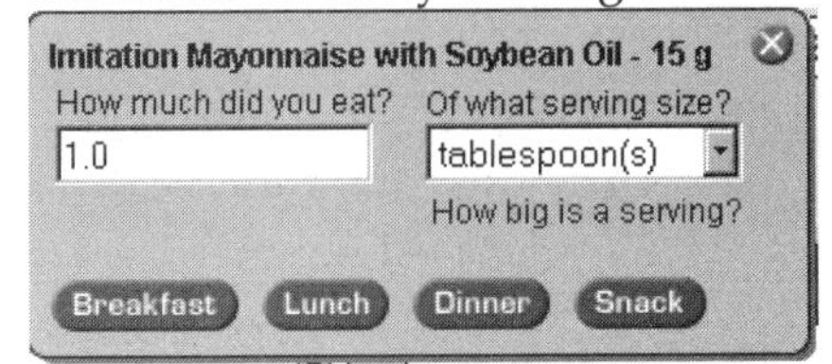

If you ate a food item that you cannot find listed in the database, but you have the label from the package with the Nutrition Facts panel, you can click the **Create a Food** button to add that food. If you ate a homemade dish prepared with foods that are already included in the database, you can click the **Create a Recipe** button to enter the ingredients, and *DA Plus* will calculate the nutrients in the recipe based on those ingredients.

When you have finished entering a day's intake, you can check the nutrition information by clicking on the **Create Reports** button. You can select which report to view by clicking on the vertical menu on the left side of the screen. You can view the following:

- Profile DRI Goals
- Macronutrient Ranges
- Fat Breakdown
- Intake vs. Goals
- Source Analysis
- Food Pyramid Analysis
- Intake Spreadsheet
- Exchanges Spreadsheet
- Activities Spreadsheet (only if you have entered your activities for each day using **Track Activity**)
- Energy Balance (only if you have entered your activities for each day using **Track Activity**)

Look at each of the reports to familiarize yourself with data presentation and interpretation.

As you complete each day's intake, be sure to check the entries over very carefully to avoid errors and therefore nonsensical data. Look especially at the total calories eaten. A reasonable amount of calories is in a range of 1500 to 3500 (depending on gender and activity level). If it is not within this range look for possible errors in your food list.

When you have entered all 3 days and have checked for accuracy, it is ready for review by your instructor.

To expedite this review, please send your login information (username and password) for *DA Plus Online* to your instructor by e-mail. Your instructor will review your food entries online and let you know if any corrections are needed or if you are ready for the Part B, the follow-up analysis.

To print reports, click on the **Print** button in the blue title bar at the top of the report window.

DIET ANALYSIS PLUS ONLINE ASSIGNMENT – PART B
Follow-Up Analysis

This follow-up analysis involves finding and transferring data and an essay, which will require a thoughtful written analysis of your diet.

- Proceed only after your instructor has reviewed your data and you have made corrections as needed.

- You will need to refer to all of your daily reports and average reports of your 3-day diet intake generated by the *Diet Analysis Plus Online*. You will also need to refer to your textbook to answer some of the questions.

- Familiarize yourself with the information on the reports.

    - Profile DRI Goals – Gives the recommended amounts based on the personal data you entered into the program – also referred to as the Dietary Reference Intakes. Please note that for some of the nutrients, the amounts recommended are amounts NOT to be exceeded. For example: saturated fat, cholesterol, and sodium.

    - For each day (referred to as Day 1, Day 2, and Day 3 in this assignment), you will have access to Macronutrient Ranges, Fat Breakdown, Intake vs. Goals, Source Analysis, Food Pyramid Analysis, Intake Spreadsheet, and Exchanges Spreadsheet reports.

    - Average reports – The program will calculate the average of the days' entries and you will have data similar to each individual day's report. Average reports can be viewed for the Macronutrient Ranges, Fat Breakdown, Intake vs. Goals, and Food Pyramid Analysis reports. To obtain the average reports, choose the date for your Day 1 under "From date" and the date for your Day 3 under "To date."

- You should look at each type of report that is provided so that you can interpret the data. For example, **Intake vs. Goals** gives the actual amount you ate (Intake), the DRI recommendation (DRI), and a percentage (Intake compared to recommended amount expressed as a percent.) Generally you should strive for 100% of goal EXCEPT for cholesterol and sodium. Only Intake is listed for nutrients with no specific DRI, such as alcohol, fat, and sugar.

- To compare your intake of the energy nutrients to the recommended distribution (45%-65% of calories from carbohydrate, 20%-35% of calories from fat, and 10%-35% of calories from protein), check **Macronutrient Ranges**. For more specific data regarding your fat intake, see the **Fat Breakdown** report. This shows the percentage of total calories you consumed that came from different types of fat: saturated, monounsaturated and polyunsaturated.

- The **Source Analysis** data will allow you to view the sources of each component in your daily intake from most to least. For example, if you want to see how much of the total fat comes from each food item you entered, you can view that data.

- The **Intake Spreadsheet** breaks down the nutrition information on each food item you entered and adjusts for the amount of food you actually ate.

- The **Food Pyramid Analysis** report places the food you ate into the pyramid guide to eating.
- The **Exchanges Spreadsheet** report provides the Quantity and Exchanges for each item.
- You will need your Intake Spreadsheet reports for Day 1,2, 3 and your average reports (submit with assignment).

To print reports:

1. Click on **Create Reports** (menu is across).
2. Select name of report, such as **Macronutrient Ranges**.
3. Choose the date for your Day 1 under "From date," and choose the date for your Day 3 under "To date." Click the **+Calculate** button.
4. Click the **Print** button in the blue title bar of the report window.
5. Repeat steps 2 through 4 for each of the other average reports: Fat Breakdown, Intake vs. Goals, and Food Pyramid Analysis.
6. Select **Intake Spreadsheet**. Select the date for your Day 1, click **+Calculate**, then click **Print**.
7. Repeat step 6 for Day 2 and Day 3 to print the Intake Spreadsheet for each day you recorded intake.

Essay portion of analysis: This will not be accepted in a handwritten format. It must be computer generated (or typewritten). The essay should be at least 2 pages in length (standard margins, double spacing). Points will be deducted for insufficient length. You are expected to demonstrate understanding of nutrition principles as they apply to your personal profile, lifestyle, and dietary practices. You must include responses to the following in your essay.

- Do you feel the recommended calories are high, low or just right for you?
- As you answer this, include information about your age, activity, occupation, health, and fitness level and any other factors that could influence your calorie needs.
- Reflect on the way your lifestyle and perhaps ethnic background or your early family food habits influence your food choices.
- How would you rate the overall quality of your diet? Include data from your 3-day intake to support your rating. It is advisable to first complete the data transfer portion of this assignment before you complete the essay portion.
- What current lifestyle factors contribute to your present diet? Examples could include lack of time, dislike or like of certain foods, interest in healthy eating, etc.
- What dietary changes would you like to achieve?
- What is your position on dietary supplements?
- If you take any dietary supplements such as a multi-vitamin and mineral supplement, what type (brand)? Are the amounts greater than the DRI?
- What are your reasons for taking the supplements?
- Based on the results of your 3-day intake, do you think supplements might be useful for you?
- How have this course and this analysis influenced your dietary practices?

by Judith S. Matheisz

Diet Analysis Plus Online Assignment Part B Worksheet

- Attach a copy of the average reports and daily Intake Spreadsheet reports. (Refer to preceding instructions before proceeding.)
- Please provide your username and password for the *DA Plus Online* program in case the instructor needs to check your reports. Remember the user name and password are case sensitive!

Username ____________________________ Password ____________________________

- Attach the essay portion of this assignment.

Refer to the Energy section of the Intake vs. Goals report and transfer your data below:

	Intake	DRI	Goal %
Calories			
Protein (g/kg/day)			
Carbohydrates			
Fat			

Does your calorie intake represent an accurate intake for you? ______________

Refer to the Macronutrient Ranges report and transfer your data below:

Source of Calories	%
Proteins	
Carbohydrates	
Fats	
Alcohol	
Total of above	

(total should equal 100% ± 1%)

- The energy nutrient guidelines offer flexibility in the mix of fats, carbohydrates, and proteins. They suggests that adults can meet their energy and nutrient needs and reduce the risk of developing chronic diseases by eating 20%-35% of calories from fats, 45%-65% from carbohydrates, and 10%-35% from protein. Comment on your % calorie distribution according to the guidelines.

- Refer to each day's Intake Spreadsheet and look at the column heading Carb (g). Sort your carbohydrate food sources as to the type of carbohydrate (complex, simple nutritious, concentrated). Only list foods that actually provide at least 15 grams of carbohydrate. List 3 per day.

Complex carbohydrate (mostly starch & fiber rich foods)		
Day 1	Day 2	Day 3

Simple nutritious carbohydrate (fruits, milk, and milk products)		
Day 1	Day 2	Day 3

Simple non-nutritious carbohydrate (sugary foods such as candy, soda pop, desserts)		
Day 1	Day 2	Day 3

- Which type of carbohydrate is present in the greatest amount?

- How does it compare to the recommendation that most of the carbohydrate in the diet should come from complex carbs, followed by simple nutritious, and sparingly from concentrated sugars?

Refer to the Carbohydrates section of the Intake vs. Goals report and transfer your data below:

	Grams	Goal %
Dietary Fiber		

- If you did not meet 100% of goal, list at least five (5) specific foods which are fiber rich that you would include in your diet to improve your fiber intake.

Refer to the Energy and Fat sections of the Intake vs. Goals report and the Fat Breakdown report, and transfer your data below:

	Grams	% of Total kcal (except for Fat)
Fat		Within DRI range? ☐ yes ☐no
Saturated Fat		
Mono Fat		
Poly Fat		
Cholesterol		

Refer to the Fat Breakdown report for Source of Fat and transfer your data:

	%
Saturated Fat	
Mono Fat	
Poly Fat	
Other/Unspecified	
Total of above	

(should be the same as Fat on Macronutrient Ranges report)

- What type of fat is likely to be in the Other/Unspecified category?

- Refer to each day's Intake Spreadsheet and look at the column heading Protein (g). Categorize your protein food sources into those that are mostly from animal foods and those that are mostly from plant foods. List 2 per day from each category.

Plant food/protein grams		
Day 1	Day 2	Day 3

Animal food/protein grams		
Day 1	Day 2	Day 3

- Which category (plant or animal) provides most of the protein grams in your diet?

- Are the protein food sources in your diet also providing a lot of fat? If so, give some specific examples.

- What changes would you like to make in regard to your protein intake?

Refer to the Intake vs. Goals and transfer your data below:

- List the vitamins and the goal % which are equal to or greater than 50% of DRI.

Vitamin	Goal %	Vitamin	Goal %

- List the vitamins and the goal % which are less than 50% of DRI.

Vitamin	Goal %	Vitamin	Goal %

- For each vitamin which is less than 50% of the DRI, research specific foods (refer to your textbook) rich in the vitamin and list foods you would include in your diet to improve your intake.

- List the minerals and the **goal** % which are equal or greater than 50% of DRI.

Mineral	Goal %	Mineral	Goal %

- List the minerals and the goal % which are less than 50% of DRI.

Mineral	Goal %	Mineral	Goal %

- For each mineral which is less than 50% of the DRI, research specific foods (refer to textbook) rich in the mineral and list foods you would include in your diet to improve your intake.

Refer to each day's Intake Spreadsheet, look at the column heading Alcohol (g), and transfer your data.

	Alcohol grams
Day 1	
Day 2	
Day 3	

- What are the sources of alcohol in your diet? Is this representative of your usual intake?

- What changes would you like to make in regard to alcohol intake?

Refer to the Food Pyramid Analysis report and transfer your data below:

	Recommended servings	Servings consumed
Fats, Oils & Sweets		
Milk, Yogurt & Cheese		
Poultry, Fish, Dry Beans, Eggs & Nuts		
Fruits		
Vegetables		
Bread, Cereal, Rice & Pasta		

- What changes would you like to make regarding your distribution of servings in the pyramid?

- What foods contributed to your fats, oils & sweets servings? When interpreting your results for this category, refer back to your actual data. If you did not exceed your fat limit and you do not eat a lot of sugary foods, the number of servings may not be as excessive as you think.